From the time they're lit[illegible] about their perfect wedding

From the tim[illegible] ream about their perfect

From the tim[illegible] ream about their perfect

From the time they're little, girls dream about their perfect

[illegible] the time they're little, girls dream a

From the time they're little, girls dream a

From the time they're little, girls dream a

From the time they're [illegible] eam a

家族荣耀 豪宅铭刻

Family Glory by Individual Mansion

（上）

open 欧朋文化 策划

黄滢 马勇 主编

华中科技大学出版社
http://www.hustp.com
中国·武汉

前言
Preface

优选一座府邸，期许美好未来

在中国古代，一个兴旺发达的家族总是要造一座祖屋，为后代子孙聚居建一个遮风避雨，繁衍生息的场所，后代的子孙有出息了，无论是做了官还是发了财，或者成为一方巨擘，都会修缮或扩建祖屋，以维护这个家族的生生不息，并作为家族未来的根基或预留的退路。这样的祖屋在山西、福建、江西、安徽、江苏、浙江、云南等地还有一些留存。中国非常重视血脉的繁衍，历史悠久的家族有时比一个朝代屹立的时间更为久远。因此有些家族的祖屋是非常可观的，规模大的可达千亩，街道相接，院落相连，雕梁画栋、富丽堂皇，内部不但有一套套居住的院落，有的连家祠、书院、戏台、街区、店铺、花园都配备齐全，走进去如同进入一座小城。当然这样的祖屋绝非一朝一夕建立起来的，往往是创始人开个头，子子孙孙数百年扎根繁衍，将其不断扩容兴建而成，这样的祖屋往往需要建造数百年，才有那样的规模。比如山西的王家大院、常家庄园，东阳卢宅等（具体参见本社出版的《中国最美的深宅大院》（1~3））。祖屋不只是生活的地方，也是一个家族举族庆典，族裔团聚，传承家族文化的所在。祖屋不仅仅只是一座房屋，更是家族血脉凝聚的见证，家族历史的见证，家族荣耀的见证。

在国外，这样的家族式庄园也非常多，比如对文艺复兴产生深远影响的美第奇府邸、罗斯柴尔德家族的庄园、霍亨索伦家族城堡等等，这些家族庄园有的成为公共景点，有的还为家族所有，成为家族的标志与骄傲。

当今社会，随着时代的发展，现代家庭日渐小型化，儿女长大成家了，大都慢慢从家庭分离出去，各自成家，几代同堂的家庭越来越少，同族之间聚居的更是少见。现代成功人士置业，满足的主要还是自行居住和会见宾客、接待亲友等需求，为家族的聚居而购置的非常少见。但是作为一位成功者，或者一个家族崛起的创始人，一定希望自己的成就、经验、人生感悟能够向后代传续下去，这需要一个有形的东西来收纳或者铭刻。一栋足以传承后世的宅邸无疑是上佳选择。既有居住的实用价值，也能在招待亲友时倍添光彩，还可以陈列创始人的珍藏，更有良好的保值价值，哪怕为后人准备一笔资产，为家族复兴留下一片基业，或者给子孙团聚提供一处值得留念的场地，也需要有一所宅邸来承载这一切。这样的宅邸兼具豪宅的功能和基业的象征，比一般的豪宅更上一层，或许我们可以用“府邸”来称之。

如果说豪宅是“为满足社会富裕群体对稀缺资源较多占有的一种欲望而定制的豪华住宅”，那么府邸应是“建筑艺术与文化沉淀相结合，具有家族传承价值的豪宅”。有些人或许有多套豪宅，甚至在世界各地置产，但是能成为府邸的只是其中一栋。

美国第一共和银行曾对豪宅定下如下标准：建筑面积为 300~600 平方米、有 3~6 间卧室和 3~6 个卫生间，并且价值在 100 万美元以上的地产。这个标准也不是绝对的，不同国家不同地区，经济发展水平是不一样的，豪宅更应该是一个相对的概念，只要比周边物业高出一定档次，就可以视为该地区的豪宅。成为豪宅一般要满足以下几大要求：

1. 对稀缺资源的占有。豪宅的珍贵性常常体现在土地的稀缺价值、特殊资源价值、景观价值、建筑特色和城市的人文价值上，如特殊的海景资源、湖景资源、河景资源、山景资源、森林资源、区域在城市中的不可替代的人文资源等等，例如法国地中海沿岸、美国纽约中央公园、日本东京湾等。

2. 豪宅的低密度性和舒适性。不是面积大就算是豪宅。它需要在一个优美

安全私密的环境里，最大化地占有优质资源，享受全方面的舒适性。低密度要求每户均拥有足够的绿化面积，开阔的尺寸，适当的距离。

3. 完善而高品质的配套设施。没有完善的配套设施，就没有高品质的豪宅生活。富豪们除了居住的需求，还要满足休闲、娱乐、健身、社交等需求。这就要多样化的配套设施来满足，比如会客大厅、恒温泳池、健身会馆、私家花园、交际会所等等。红酒吧、咖啡吧、阅读书吧、茶吧的配套都属常规，有些豪宅区甚至配备了小型教堂、温泉会所、私家音乐厅、壁球馆、小型剧场等等，为了豪宅的生活品质可谓不遗余力。

4. 高质高效尊贵的物业服务。配套再高端，生活还是与人打交道，因此还需要高质量的服务来完善和维系。五星级酒店式服务只是基础，尊贵的体验还在于有人悉心周到地满足社区生活所需，有特别需要时也有专人为之倾力服务。

5. 豪宅的优设计与高品位。任何一座闻名于世的豪宅都有著名的设计大师为之精心设计。从建筑、景观到室内，设计师用他们的聪明才智来提升豪宅的品位与格调，使之诞生之初就能光彩照人，并且历经时光的考验仍能保持优雅高贵的姿态。豪宅没有深厚的美学基础打底，就成了暴发户炫耀的笑话了。

6. 豪宅适度超前与可持续性。与时俱进也适用于豪宅的建造、配套与服务。科学技术的发展，材料的更新迭代、工艺的进步发展，都为豪宅的舒适、便利、尊贵提供了更多的可能。豪宅本来也应具备引领潮流的特质，创新也是推动豪宅进步的动力。比如比尔·盖茨那样的科技之家，可以用网络传感系统自动调节阳光的强度，空气的温度、湿度……

7. 豪宅的私密性与安全性。所有豪宅都尽可能地在公众视线之外，为了保证安全和私密性，豪宅周边一般都有高墙围绕，配备的监控、防护、报警系统也严于一般住宅。在世界各地的豪宅区中，无论是美国的比华利山，还是新加坡香格饭店周边的豪宅区域，均坐落于“路的尽端”，以远远避开街道或公路两侧及一切视线能及的地方，这是一种通常的规划手法。

相对豪宅的硬件指标，我们认为值得传承的“府邸”应该还有一定的软件

要求：

1. 文化艺术的沉淀。

在中国古代，有的豪门士族曾经比皇帝还要尊贵有权势，比如东晋及六朝显赫一时的高门士族琅琊王氏、陈郡谢氏以及博陵崔氏。还有的家族在中国血雨腥风的朝代更替中延续了两千多年，如山东曲阜的孔府。国外也有延续了几百年的豪门世族，如罗斯柴尔德家族、维特根斯坦家族，当他们被称为“欧洲第六帝国”的时候，美国还没有建国，是全球最神秘最古老的家族；还有霍亨索伦家族，曾是勃兰登堡、普鲁士及德意志帝国的统治家族；此外维特根斯坦家族，也是欧洲最显赫的豪门家族之一，有着数百年的悠久历史，家族的产业遍及钢铁、铁路、轮胎、金融和建筑业，对整个世界的影响涉及政治、经济、文化、哲学等诸多领域，拥有“钢铁帝国”之称。

为什么有的家族能延续辉煌，而有的却式微或湮灭了呢。这其中去除了运气的成分，跟家族的文化传承，子孙培养有莫大的关系。做人原则、思想品德、眼界视野、行事守则、知识水平、创富能力、谋略策划、执行能力、守成手腕等皆是文化，以人文思想为后代镀金，才是最宝贵的财富。世界上最经久流传的顶级豪宅，即便其奢华可以被复制，但其历史沉淀所产生的人文风光，却是无法被超越的。而人文艺术的沉淀，绝非一朝一夕能够完成，它应该被渗透到生活中的各个环节，在日常生活中有意无意地培养与提升。在中国传统深宅大院中，遍布的石雕木刻、玉雕彩绘、楹联碑文都可以看出，有远见的祖先们对子孙后代的教育是极其重视的，将理念融入生活的方方面面。国外的顶级庄园、城堡也重视文化艺术的熏陶与影响，很多家族在扶植艺术创作上不遗余力，如美第奇家族、贝利尼家族。当积累了大量的财富后，收藏艺术品，赞助艺术家成了西方历史悠久的文化生态之一。

相比大陆豪宅在硬件、名牌设施上的大笔投入，台湾豪宅的意识就颇值一赞。在台湾，所推的豪宅若是没有几件艺术品陈列出来，简直要被人当土豪嘲笑。在台中有片区域，名为七期市地重划区，围绕着市政府，豪宅林立，家家都有艺术品展示，有的是找名家定制，有的是在画廊买回，有的是从市场上拍卖回来的，豪宅级别越高，陈列的艺术品规格也越高，行走其间，如同巡视露天艺术画廊。还有的豪宅在售楼期间，将售楼部布置成主题艺术展，给受众参观；有的豪宅在交楼后，每隔一段时间，就将公设内的艺术品更换一批，持续不断地提升住户的审美能力。当然，以上这些属人为地创造文化氛围，我们还有一个先天之利，那就是五千年文明积累下来的历史文化，若将这些先天之利与后天的艺术培养结合起来，相信定能为豪宅持久价值提供深厚支持。

2. 重视设计。

优秀的设计不只让豪宅更美观，更好用，更安全。真正富有洞察力、创造力和远见的设计，不但能拔高豪宅的形象，还能持续地为豪宅升值加分。比如，一个形象鲜明的豪宅，能够成为一个区域的标志，进而巩固和提升豪宅的地位。还有优秀的景观规划，能够从水文、地理、风向、节气等多方面进行考虑，使建筑在布局上更为生态化，让住区内夏季凉爽，冬季温暖，极端气候状态下也能更安全可靠，并实现健康居住，环保生活。还有就是住宅的尊贵性、文化性与舒适性，都能通过设计的手段推向极致。所以挑选能够成为“府邸”的豪宅，并不是地段好、品牌优、价格高就够了，好的设计能让其价值倍增。

3. 前瞻的目光。

购买豪宅本身就是一项投资，而成为府邸要求有更前瞻的目光。这份资产能不能长久地保存流传，以及周边环境能否持续向好，可持续发展的潜力够不够，都是要业主悉心考量的指标。至于选择的这座府邸是别墅还是平层，其实并不重要，比如伦敦的海德公园一号，尊为全球最贵的平层豪宅，它的居住舒适性和配套的确堪称顶级，社区配套设施非常豪华，包括公共温泉和壁球场等。公寓户型设计由著名设计师理查德·罗杰斯亲自操刀，智能安全系统是公寓最大亮点。还有香港的天玺，2009 年“至高无上”的叫价高达 54 万元 / 平方米，由两座约 270 米高的摩天大厦组成，成为香港最高的住宅，配设私人泳池及 186 平方米的空中花园，对于“城市客厅”式平层住宅的打造力度之大，令人折服。府邸可以从适合当下阶段最好的选择开始，慢慢淘换。

选一座府邸是对美好前景的期许，是对未来家族发展的规划，同时是人生成就的印证。志存高远，放眼未来，是挑选一座府邸的意义所在。

目录
Contents

精装豪宅
Well-decorated Mansion

阿拉伯豪宅
Arab Mansion

生态大宅
Eco-Mansion

精装豪宅

Well-decorated Mansion

Mysterious Charm to Interpret Authentic British Style

神秘魅力绽放，演绎纯正奢华的英伦格调

项目名称：中星红庐别墅（英式）
设计公司：鼎族设计
设计师：吴军宏
摄影：三像摄建筑摄影机构 张静
面积：942.8 m²

Project Name: Red Villa (British)
Design Company: Prosperous Clan Adorn Design
Designer: Wu Junhong
Photography: Threeimage Zhang Jing
Area: 942.8 m²

本案定位为英伦风格，面积为 942.8 平方米。针对当下中国高端富裕阶层大多向往欧洲以前的贵族生活，设计采用了大量深色的护墙板、不同柱式的造型和各种花式的线条以及小装饰，来打造一套纯正的英式风格别墅，力图把人们带回 19 世纪的欧洲，使他们充分感受到英国贵族的从容、淡定和雍容华贵。

从别墅玄关处步入，玄关以低调的华丽之姿尽显内敛的贵族气质，沉稳的空间追忆出 19 世纪老宅的复古印象，回转的楼梯不仅使挑高的两层楼过道看起来不会显得过于空旷，也丰富了空间的层次感和维度。别墅一层主要以客厅、餐厅、起居室为主要功能，同时也设计了一间套房作为父母房，方便使用。早餐厅、中餐厅和西餐厅围绕着西式厨房，让三者的空间更为互动。西餐厅强调强烈的秩序感，体现出大宅风范。一整排欧式繁复的花样吊灯，充满隆重的仪式感。层层帘幔取代了实体门，不但令空间通透，也带来舞台般的戏剧美感。客厅格局方正大气，布局体现出古典主义对称、平衡的美学特征，精美的雕花壁炉是空间的视觉焦点。体积敦实、造型端正的家具，营造出一种怀旧的古老气息。设计在客厅边上挖出一块空间作为起居室，既让功能得到满足，也使整体空间更为宽敞、气派。

二层对主卧的更衣室作了调整，使主卧空间更为完整、开阔。另对主卫也作了调节，让空间更为豪华，更具套房概念。卧室与书房为开放式格局，一侧的英式雕花书桌椅令人联想到手拿烟斗的“老克勤”。卫生间同样延续空间的怀旧主题，镜子的设计颇具匠心，将“圆”与“方”巧妙结合。

地下室以休闲娱乐功能为主，内含桌球区、视听区、酒吧区、酒窖区、雪茄区、棋牌室和水疗健身区。酒吧区浓烈的暗红色 Art Deco 灯具与深褐色空间搭配，产生强烈的装饰性。地下室调整了视听室的位置，充分利用了下沉庭院的采光和空气流通，让地下室空间更为宽敞、明亮。而原水疗区则增加了两级踏步，使水疗池和跳操瑜伽房相结合，功能更加完善。

中星红庐别墅，带给人们的是视觉上的震撼，当你走进门庭，恍若置身于 19 世纪西欧的古老城堡，高贵而神秘。钢琴键般的旋转楼梯、宫殿式的玄关、教皇实木落地柜以及骑士壁画等，烘托出令人惊叹的皇家气息；璀璨的吊灯，发出富贵的金色，使整幢别墅贵气逼人；

特有的廊柱设计，沿袭了欧洲文艺复兴时期的华贵、典雅；精美的雕花、闪亮的银器、温暖的壁炉、美丽的帘幔，连餐厅都是那么的奢华和优雅。置身于此，你会不由自主地浮想联翩，想象自己身着华衣锦服，穿梭于名流之间，受到众星捧月般的拥戴，歌舞笙箫，满目琳琅，享受“此曲只应天上有，人间能得几回闻”的奢华生活。

DANIEL LIBESKIND JÜDISCHES MUSEUM BERLIN

Dignity and Taste in Low-key Luxury

奢华内敛中彰显尊贵品位

项目名称：法式扭卷
项目位置：加利福尼亚
建筑公司：芬藤建筑
设计公司：兰德里设计
设计师：琼·本克
面积：3 906 m²

Project Name: French Twist
Location: Beverly Hills, California
Construction Company: Finton Construction
Design Company: Landry Design Group
Designer: Joan Behnke
Area: 3,906 m²

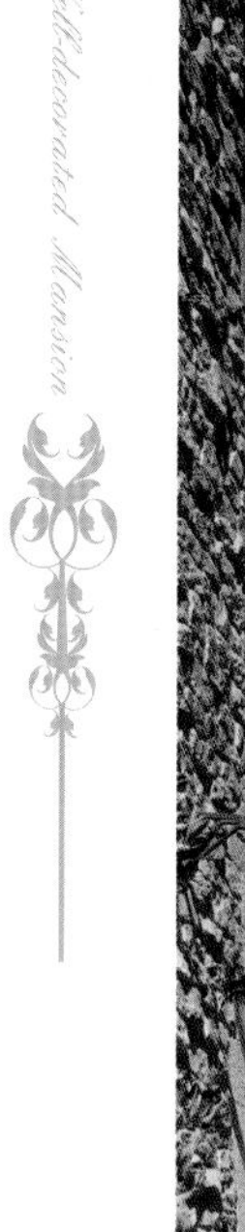

比佛利山庄，10 117 平方米的地基，便是本案 3 716 平方米的超大公馆。空间设计尽力避免因开阔而带来的僵硬感，以其人性化的比例予人一种内敛的奢华。

玄关直入复合娱乐区域，那里有餐厅、家庭室、客厅等等。餐厅里遍布古董家具、装饰器物。其间铺陈年代久远，跨越多个世纪，空气中因此弥漫着一种巴黎式的公寓感。中世纪的法式羊皮铜绿烛台，精致的模具，定制的古董家具饰面衬托着的灰色墙体，以及海马毛、丝绸等其他质感的面料应有尽有。家庭室里，装有横梁的天花、暖色调的调色板、休闲式的家具、电视让此处有了一种乡村别墅的感觉。

凉亭是主人最喜爱的地方之一。一如其他空间，一专多能。从客厅或图书室都可以轻松到达。两面墙体镶有折叠的门，温暖时打开、寒冷时关闭。烈火燃烧的壁炉，名师手笔的衬垫、藤质家具强化着此处内外交融的主题。

卧室里的四柱床，让你安然入睡。正对着壁炉的休息室，让夜晚的阅读变得尤为舒适。沙发后的橱柜镶嵌着电视。躺在床上，视线张望，也可以轻松地看到电视。

宽阔的建筑空间让各功能分区有了进一步的细化，数目繁多，但不紊乱。有了有氧运动及地秤，家居生活的空间则很容易地变成了一个温馨的健身房。

This 3,716 square-meters residence, located on a 10,117 square meters lot in Beverly Hills, voids the rigid formality often associated with mammoth dimensions, offering instead human proportions and a restrained sense of luxury.

Directly off the entry are areas for entertaining, including the dining room, the family room, and the living room. Designed to evoke the ambience of a Parisian apartment, the living room displays a collection of antique furnishings and decorative pieces that spans centuries, giving the impression of having been accumulated over time. Mid-century French parchment-and-brass scones, intricate moldings, grey walls in a custom antiqued furniture finish, and a variety of textured fabrics-including mohair, silk, and horsehair, which covers an ottoman-define the space. The family room, with its beamed ceilings, warmer color palette, casual furnishings, and television, has the feel of a country house.

The loggia is one of the homeowner's favorite areas and one of the residence's most versatile ones. Accessed from either the living room or the library, the loggia has folding doors that open up two of its walls to the outdoors on warm days and enclose the space entire during the colder months. The fireplace provides warmth, while the cushioned wicker outdoor furniture from Brenda Antin in Los Angeles accentuates the indoor-outdoor theme.

In the master bedroom, a canopy bed offers a cocoon like retreat for sleeping, while the sitting area, which faces a fireplace, creates a cozy enclave for reading in the evening. A cabinet behind the sofa conceals a television that can be watched from the bed.

The home's ample square footage allows for a variety of specialized spaces, all of them used regularly. Equipped with variety of cardio and weight machines, the home's fitness center makes getting to the gym easy.

LUXURY TOYS

Dignity Comes from History

历史成就尊贵，优雅代代传承

项目名称：大吕克城堡
原建筑师：Mattieu de Bayeaux
设计公司：Timothy Corrigan, Inc.
设计师：Timothy Corrigan
客户：Timothy Corrigan
地点：法国卢瓦尔河谷
面积：52 666.66 m^2
（主城堡占地面积：2 322.576 m^2）
用材：法国石灰石

Project Name: Chateau du Grand-Luce
Original Architect: Mattieu de Bayeaux
Design Company: Timothy Corrigan, Inc.
Designer: Timothy Corrigan
Client: Timothy Corrigan
Location: Loire Valley, France
Area: 52,666.66 m^2
(Floor Area of Main Chateau: 2,322.576 m^2)
Materials: French Limestone

大吕克城堡展现了法国启蒙运动时期的一种最珍贵的建筑元素。

——法国著名设计师B. Chauffert-Yvart

历史

大吕克城堡历史悠久，其历史可以追溯到18世纪。这座城堡是由设计师Mathieu de Bayeux于1760—1764年年间为大吕克男爵Jacques Pineau Viennay设计的。经过几十年的精心设计和五年多持续不断的建造，Pineau Viennay本以为终于能见到自己的杰作，却未曾料想，在进门时由于过度激动而导致心脏病发，从而，抱憾而终。因此也更未见到花园内国王路易十五世赠送的乔迁之礼——众多雕像（确切地说，这些雕像是凡尔赛宫内雕像的复制品）。

大吕克城堡曾经接待过很多贵客，比如启蒙运动中的杰出人物代表伏尔泰、狄德罗和卢梭，甚至德国著名作家歌德也在此逗留过，歌德第一次见到Pineau de Viennay的时候还在斯特拉堡求学。城堡最终由男爵的女儿Louise Pineau Viennay小姐继承。1781年，城堡附近的村庄发生火灾，烧毁了144间房屋。Louise Pineau Viennay小姐心地善良，打开城堡大门救济村民，并资助村民重建家园。为确保村庄从此免受火灾，重建的村庄采用了曾经用于建造城堡的石头作为主要建材。几年以后，Louise Pineau Viennay小姐的善举便得到了回报。法国大革命期间，Louise Pineau Viennay小姐和城堡在村民的保护下得以幸存。城堡的室内几乎保存完好，这是法国唯一一座在法国大革命期间保存得如此完好的城堡。

这座城堡及产权200多年来一直属于同一个家族。第二次世界大战期间（1939—1945年），大吕克城堡变成了一个军事医院，成为英国军官的医疗场所。此时，来自卢浮、里尔以及法国其他博物馆的700多件珍贵的画作也收藏于此。

二战结束后，这座城堡不再归私人所有，而成为了一个医疗中心。1996年，由于该城堡是为数不多的国家标志性建筑物，法国政府决定重修。法国政府启动修缮花园这一伟大项目，并于1999年对公众开放。2005年，法国政府决定将其产权私有化，并且选中 Timothy Corrigan 作为候选人继续修缮城堡花园以及室内，使之成为私人住宅。

设计

这座城堡建于1760—1764年间，城堡以及其众多的附属建筑物（教堂、马厩、厨房、面包房、清洗房、橘园和仆人住所）都是法国国王路易十五期间的建筑风格的典型代表。城堡中轴对称，从入口到荣誉展馆延伸到城堡的北面，中轴线将城堡分为两部分，住房外部与花园相连，组成法式阳台。这座城堡是古典装饰概念和装饰图案的绝佳案例之一，采用了对称、对齐窗口、柱顶以及山形墙饰建筑装饰方法。由于建造的初衷是将城堡建造成一座夏宫，所以在城堡的南面以及豪华沙龙室都能见到许多园艺工艺（花篮、喷壶、耙子等）。

重新修缮

Timothy Corrigan 勇敢地承担起修缮大吕克城堡室内的工作，使18世纪的夏宫在三年内恢复原有的生机。Timothy 向法国政府提交购买大吕克城堡的申请时，不仅需要与开发商和酒店经营商竞争，更值得一提的是，他不仅是唯一的私人申请者，而且还是众多申请者中唯一的美国人。法国历史文物保护机构 L'Architecte des Batiments de France 称，鉴于大吕克城堡是法国启蒙运动时期的建筑风格的珍贵代表，因此不主张进行大改建。Timothy 对此看法表示赞同，最终成功购买城堡。

The chateau at Le Grand-Lucé represents one of the most precious elements of architecture of the French Enlightenment.

B. Chauffert-Yvart, Architect of the French Monuments

History

Chateau du Grand-Lucé has a rich history dating back to the 18th century. Designed by Mathieu de Bayeux for Jacques Pineau Viennay, Baron de Luce', Chateau du Grand-Lucéwas built between 1760 and 1764. After several decades of planning his new chateau and five years of continuous construction, Pineau Viennay arrived to see his finished masterpiece. Unfortunately, upon arriving at the gates of his chateau he was so overwhelmed that he died of a heart attack even before entering the chateau or seeing the numerous sculptures (exact replicas of the ones that he had at the Chateau du Versailles) that King Louis XV had placed in the gardens as house-warming gift.

Celebrated guests of the chateau have included such luminaries of the Age of Enlightenment as Voltaire, Rousseau, Diderot, and even the German writer, Goethe who had first met Pineau de Viennay while he was a student in Strasbourg. The chateau was eventually inherited by the Baron's daughter, Mademoiselle Louise Pineau Viennay. In 1781, a major fire broke out in the village surrounding the chateau and destroyed 144 houses. In a gesture of great good-will Mademoiselle opened the doors of the chateau to the village and paid to have the town rebuilt, only this time in the same stone that the chateau was built so that they would never again face the same risk of fire. Only a few years later, this act of generosity was directly repaid, when the Revolution occurred, and both Mademoiselle and the Chateau were protected by the resident's of the village. Every bit of the interior of the chateau survived in tact from this time and it is one of the only chateaux in all of France that weathered this turbulent time totally unscathed.

The chateau and its property remained in the same family for over 200 years. During the Second World War (1939-1945) the chateau served as a hospital for British Military officers. 700 important paintings (Rubens, Watteau, Fragonard, and Van Dyck) from the Louvre, Lille and other French museums were also hidden at the chateau throughout the war.

Following the war the Chateau left private hands and it became a medical center for a number of years. Then in 1996, due to the architectural importance of this Listed National Landmark, the French government decided that it should be restored for the public. An ambitious project of restoring the formal gardens began and in 1999 the gardens were opened to the public. In 2005 the government decided to return the chateau to private ownership and selected Timothy Corrigan as the ideal candidate to continue with the restoration of the gardens and to begin the interior restoration of the chateau as a private residence.

Design

Built between 1760 and 1764, the chateau and the numerous outbuildings (chapel, stables, kitchen, bakery, wash house, orangery, and servants lodgings) are a typical example of the architectural style in France under Louis XV. The estate is organized around an axe of perspective that starts atthe entranceway to the Court of Honor, continues through the northern façade of the chateau, divides the floor plan of the house in two, carries on outside of the house into the gardens to become the main alleyway in the formal French terraces. The chateau is an excellent example of the concept and decorative motifs of classical antiquity: symmetry, aligned windows, capitals, and pediments. As the chateau was intended primarily as a summer palace, there are numerous references to horticultural art (a flower basket, watering can, rake, etc) on both the southern façade as well as in the Grand Salon.

Renovation

The three year process of renovating the interiors of Chateau du Grand-Lucé was bravely undertaken by Timothy Corrigan who brought the 18th century summer palace back to life. When Timothy submitted an application to buy the Chateau du Grand-Lucé from the French government, the custodian of the building and its resplendent gardens, he went up against developers and hoteliers; he was the only private applicant, and the only American, among them. L'Architecte des Bâtiments de France, the institution that protects historic French monuments, declared that the chateau was a precious example of French Enlightenment architecture and they were in support of it not being dramatically reconfigured. Timothy shared the same vision and was thus awarded the chateau.

Search Everywhere

上穷碧落下银泉

项目名称：GL10 住宅
设计公司：台北玄武设计
设计师：黄书恒、欧阳毅、陈佑如、张铧文
软装设计：胡春惠、张禾蒂、沈颖
摄影师：赵志诚
撰文：程歆淳
面积：500 m²
用材：银狐、黑白根、镜面不锈钢、黑蕾丝木皮、银箔、金箔、进口拼花马赛克、黑白色钢烤

Project Name: GI10 Residence
Design Company: Sherwood Design
Designer: Huang Shuheng, Ouyang Yi, Chen Youru, Zhang Huawen
Upholstering: Hu Chunhui, Zhang Hedi, Shen Ying
Photographer: Zhao Zhicheng
Text: Cheng Xinchun
Area: 500 m²
Materials: Marble, Mirror Stainless Steel, Wood Veneer, Silver Foil, Gold Foil, Mosaic, Baking

本案为坐落于城市新区的宅邸，既揽有半山的绿意，又拥广场的辽阔视野。以其作为退休生活的启始，必然需要一番缜密而细腻的规划。玄武设计考虑屋主姐弟与母亲同住的实用需求，以及居住者对于美学风格的爱好，力求使艺术生活化、生活艺术化。最终择以现代巴洛克为基底，以其独有的收敛与狂放，配合玄武擅长的中西混搭——冲突美学，铺陈空间每一根轴线。

尚未进入玄关，已见一座当代艺术作品灵动而立，既巧妙掩饰了半弧形缺角，又以生动的童稚神情，为居所引入活跃的生机；右进，切入高耸柱式与圆形顶盖，视觉猛然挑高，使人豁然开朗，经典的黑白纯色打底，中置网烤定制家具，配合景泰蓝珐琅与定做琉璃，东西文化的灵活互动，为访客带来第二重震撼。

有别于玄关的单纯配色，屋主因业务所需，时有交谊与公务之需，特别需要一大气而有趣的客厅空间，活络人际关系。是故，玄武设计着重天然风光与人为艺术的调和，保留大型落地窗与沙发的间距，后者特别选用进口原版设计，呈现简练利落的现代风情。与此不同的是，中央大胆置入以艳紫、宝蓝与金黄三者交织而成的地毯，强化了简约与繁复的冲突美感，亦彰显法式皇家的堂然大度。

抬眼向上，一盏华丽的银色花朵灯灿烂夺目，使人倍感震撼，这个取材自苗族银饰的大型艺术品，为玄武设计与当代艺术家席时斌共同创作，外围化用鸢尾花意象，曲折花饰包复核心，间隙镶嵌彩色琉璃，使打底的银灰色更显时尚，每当开关按下，艺术品外围即有五彩灯光流转，可应不同情境而切换，上缀羽饰的大型银环绕着核心缓缓移动，隐喻着天文学——恒星与行星的概念，呈现着自然与人文的灵动对话。

穿越廊道，可进入屋主的阅读空间。两处各以深、浅色为底，再各自于细微之处缀以相反的色彩进行诠释。如，主卧书房一方面延续着公共空间的半圆形语汇，引导访客进入皮质沙发、深色书柜、石材拼花共构的豪气场域，却相当跳跃的使用清淡色泽织毯，大幅提升空间的律动感；主卧的书房，则纳入半户外的开阔设计，以白色底板铺底，却照样使用黑色书柜与铁灰沙发，抢眼的小号造型灯具，具体而微的体现了屋主喜好，展现内外呼应的生活态度。

因应屋主对于公私界线的看重，玄武设计亦将此概念纳入考虑，公共区域的门扉使用白色，予人亲近、纯净之感；进入私人区则以黑色区隔，带有隔绝、凸显正式的意义。进入次要空间，棋牌室与餐厅分据左右，二者均以白色为主调，黑白格地板、经典款水晶灯，搭配巴洛克花纹座椅、鸽灰抱枕，远观近看，各有深韵。

为使主客起卧舒适，主客卧房采用一贯的轻柔色泽，再以方向不同的线条勾勒空间表情，如主卧简练的长形线板与金黄床褥、浅蓝地毯相映成趣，减少过度堆栈的冗赘感；其余卧房则以湖水绿、天空蓝为点缀，在纯白、浅灰的基调里，窗帘、床褥与地毯稍有呈现，与牡丹纹床背板的繁复，共谱出屋主悠闲淡雅的生活情趣。

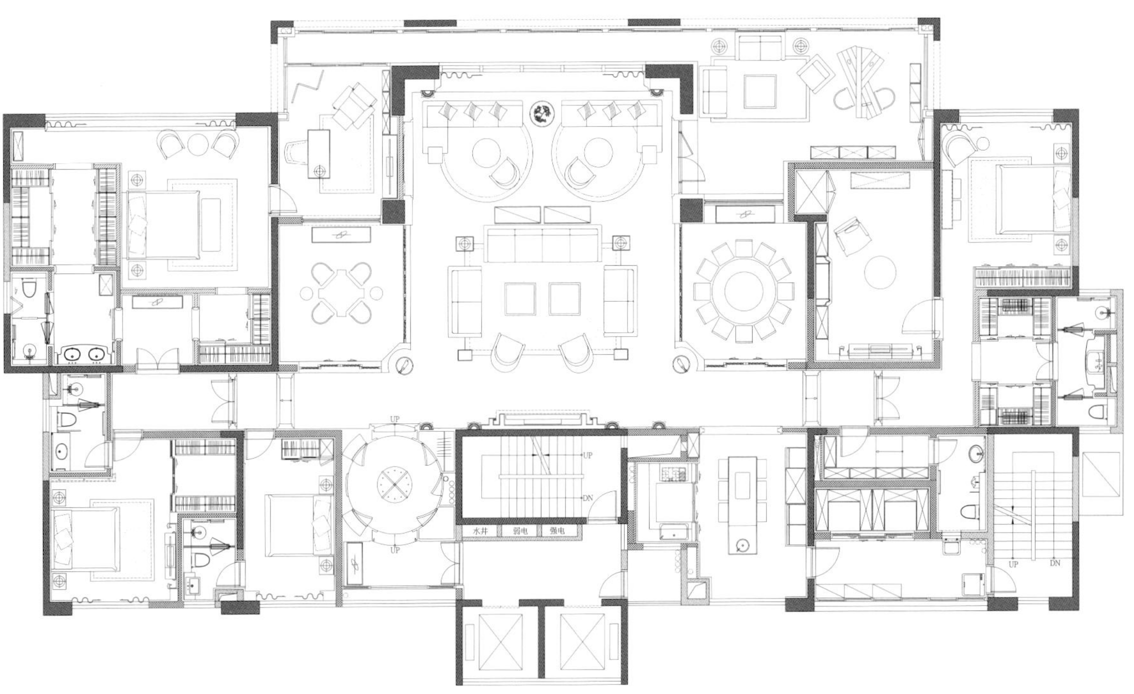

平面图 / Site Plan

A project this space is located in the new area of Taip which enjoys a setting of greenery hillside. The view of wide field through the landing window in the living roo is stressed in design. The consideration taken into of t sister and the brother living with their mother and person aesthetics of art intended for life and life destined for a leads to the Baroque base and a collision of the west a the east to lay a solid foundation for the space. The sty is conservative and unrestrained. And the approach is t practice Sherwood is usually good at.

Just before the foyer, a modern sculpture stands vivid a lively, an artistic item not only cover the shortage of the hal arched unfilled corner but instilling vigor and vitality wi child-like expression. Along the right, the towering colum and the dome heightening the visual effect. People the feel enlightened suddenly. In a black-white tone, custo furnishings with cloisonne enamel and custom glass allow for another impact.

Differentiated from the vestibule, the living room to mee personal requirements of social life and office is done gran and interesting. The Sherwood stresses the harmony betwee landscape and artificial art to keep a big gap betwee the large landing window and the sofa. The sofa's desig involves imported original edition, looking neat, crisp an modern. The center, on the contrary, is paved with carpet o purple, sapphire blue and gold yellow, which strengthens th conflicting aesthetics of the simplicity and the complicatio while bringing out grandeur and generosity of French roya family.

Down the ceiling is a flower shining silver and imposin a shocking impact, the large artistic item inspired by idea of Miao Nationality and done by Sherwood and Xi Shibin a modern artist. Around the flower is image of flower-de luce, about its center is zigzag floriation with stained glas in the interval, which makes the silver-grey backgroun more fashion. With switch turned on, lighting of five colors go around the item, its appearance changing with mood The huge silver ring embellished with feather, making a metaphor of star and planet. This presents a lither dialogue between man and nature.

At the end of the corridor is the reading area, a space with light and dark as its theme hue and contrasting colors applied in details. The study in the master bedroom continues the semi-circle language of the public space, leading visitors onto the leather sofa. Dark book cabinet and marble parquet are unexpectedly set off with light-colored carpet, which upgrades the spatial rhyme to a large scale. As for the study in the secondary bedroom is semi-open, whose flooring is white and where the bookshelf and the sofa of the same color with those in the master bedroom, and the small lighting concretely reveals personal taste and a life attitude that the *internal and the external are echoing.*

The significance laid on the partition of the public and private spaces is embodied. Doors in the public space is white, a color generating feelings of affinity and purity. The private is shaped with black, meaning separation to highlight the formal sense. In the secondary space, on the right and on the left are respectively the chess-card room and the dining room, both of which toned with white, are fixed with black and white checkered flooring, classical crystal lamp, baroque chairs, and pigeon-grey cushions. Whether looked near or far away, each and every item looks unique and aesthetic.

In order to boast comfort, the bedrooms of the host and the guest uses the usual soft hues to sketch spatial expression with lines that extends into different directions. The long wire board in the master bedroom is concise and contrasting finely with the gold bedding and the light blue carpet to reduce the verbose sense by accumulation over much. Meanwhile, the rest bedrooms involves lake green and sky blue to intersperse in a tone of pure white and light grey. Curtain, bedding and carpet come second, but bring out a leisure and refined life with the peony-patterned bed backboard.

Globally-Charmed Clusters of, Dreamy Residence

多国风情汇聚，梦幻般的住宅

项目名称：安达卢西亚大公馆
设计师：理查德·兰德里

Project Name: Andalusian Residence
Designer: Richard Landry

安达卢西亚大公馆如同“谜箱”。虽然直面街道，但借助设计师之手，空间不再仅仅是墙体、门户之后的一个几何量体，而是以丰富的层次佑护着居家生活的隐私。

早餐厅、餐厅毗邻厨房。按本案之设计理念，正餐厅应有多种功能，其融图书室、书房为一体，用餐、阅读两相宜。拱形的天花以手工刻绘的栅条作为装饰，予人一种视觉上的趣味。黑色亮木的嵌入式书柜，给人一种历史感与温馨感。

红木的楼梯直入二楼“画廊”空间。楼上各空间无缝对接。舒适的廊道，拱顶、立柱，引领着主卧空间。赤褐色的棋盘地板、织锦的枕头延续着“摩洛哥”的主题。

空间后部另有凉亭。站在覆盖有天顶的凉亭空间，可放眼直观不远处的泳池。摩洛哥的灯笼点燃着此处的激情，西班牙进口古董瓦面装饰的壁炉提升着空间的温度，17 世纪的铸铁罩套构件，装点着空间的天花。

别样的精彩，一如 20 世纪 20 年代的好莱坞建筑质感，又如同一部无声的电影，不再有新创的异域风情，但却有浓郁的怀旧色彩。

The house is something of a puzzle box. Apparently open to the street, Landry's design offers cleverly concealed layers of privacy that are far more subtle than merely situating the house behind walls and gates.

The breakfast room and dining room adjoin the kitchen. In Richard Landry's experience, formal dining rooms are under-used spaces in many homes; to prevent this, he turned the dining room into a combination dining room/library/study, with a vaulted ceiling with hand-stenciled ribs to create visual interest, and built-in bookshelves of dark wood, to add a sense of history and warmth.

A mahogany-railed staircase leads to the second-floor gallery, and all upstairs rooms flow seamlessly off of it. A cozy vestibule, framed by arches and columns, leads to the master bedroom. In keeping with the Moroccan theme, it features a checkerboard terra cotta floor and a banquette piled high with tapestry pillows.

A covered loggia at the rear of the house, overlooking the swimming pool, is lit by Moroccan lanterns and warmed by a plaster fireplace framed by an antique ceramic tile surround imported from Spain, and topped by a museum quality, 17th-century wrought iron mantle piece.

The house is a whimsical fantasy, reminiscent of Hollywood architecture of the 1920s. Indeed, it is often mistaken for a piece of silent film history, a beautifully preserved piece of nostalgic glamour, instead of a newly created exotic fantasia.

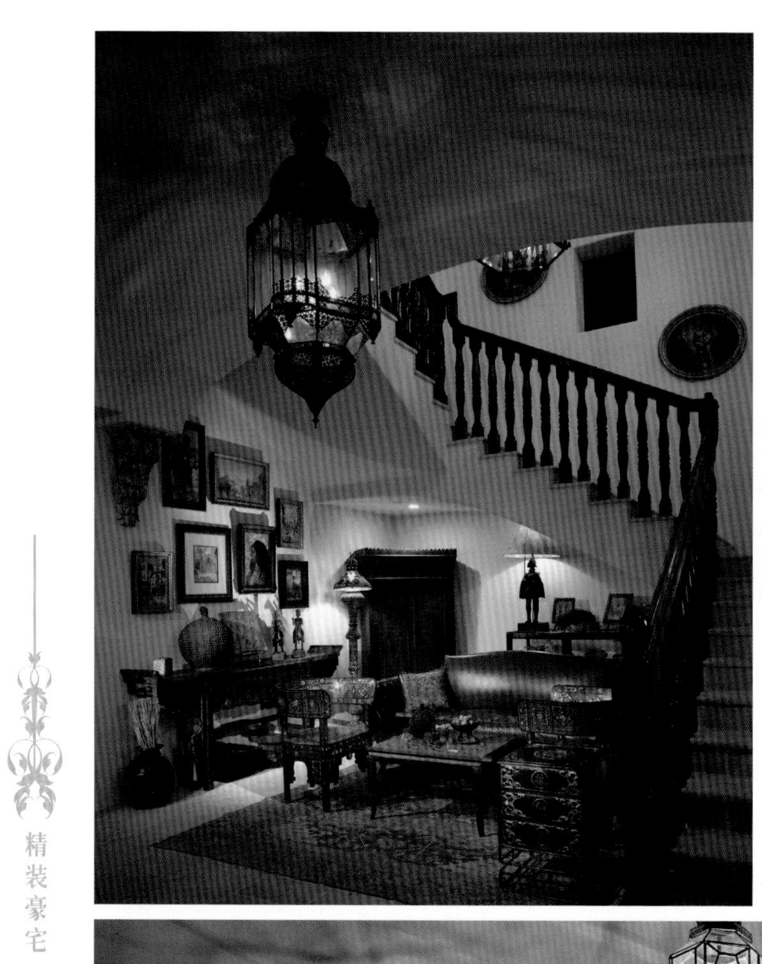

A Western City of Suzhou, Aria by Sea Blue

西方苏州城，海蓝深咏

作品名称：苏州水岸西式秀墅
设计公司：玄武设计
设计师：黄书恒、苏幼君
软装设计：杨惠涵、张禾蒂、沈颖
摄影师：王基守
撰文：程歆淳
面积：344 m²
用材：蒙马特灰大理石、原色油面崖豆木地板、梧桐喷砂实木拼、白色钢琴烤漆、明镜、灰镜、蓝色油性平光漆

Project Name: Beautiful Villa, Water Shore (Western Style), Suzhou
Design Company: Sherwood Design
Designer: Huang Shuheng, Su Youjun
Upholstering: Yang Huihan, Zhang Hedi, Shen Ying
Photographer: Wang Jishou
Text: Cheng Xinchun
Area: 344 m²
Materials: Marble, Flooring, Solid Wood Parquet, Piano Stoving Varnish, Mirror, Flat Enamel

当时间静止，风景凝结于旅人的视野，唯一抹海蓝自中心漫开，铺就整座城市的底蕴。

假如世界移形换影，将水都威尼斯迁移至中国，会是如何风景？分据世界东西的两座城市，同样对“水”有着奇异的想象。玄武设计化用威尼斯的碧蓝天色与湛蓝海洋意象，配以柔和婉约的维多利亚风，使居住者于深浅变换、线条起伏之间，体尝专属本案的深邃风情。

踏入玄关，大幅沉稳碧色沉淀着访客心绪，仿佛进入高潮之前的低沉乐音，诱人缓步轻移。步入大厅，可见西厨吧台区域的纯白色与铁灰镜面，体现强烈的戏剧张力。为扩展景深、消解低梁带来的压迫感，设计师特别利用三座连拱的流畅弧线，借由动态的视觉起伏，纾解空间压力。餐厅利用餐椅与灯具，营造素白、浅灰与碧蓝的色彩游戏，侧边一只复古壁炉，呈现精致的英伦风韵，与客厅电视墙合而为一的精巧设计，亦可见营造焦点、避免对象散乱的匠心。

延续设计主轴，设计师选用深色木皮为楼梯处主色，并于接缝处，嵌入一盏盏小型LED灯，光线影影绰绰，随着步履忽隐忽现，仿若大隐于苍穹的点点星光，为阒然的壁面点染几分趣味，也削减了色彩过于沉重的疑虑。上至二楼，大面锻铁金漆扶手蕴藏浓厚西式韵味，与实木地板、水晶灯等，维系着空间的大气磅礴。主卧，静默于一片隽蓝之中，设计师运用中国青花瓷、湖水绿等色泽，创造多层次视觉变化；次卧铺就小碎花壁纸，床背板以铆钉排出流利图腾，一如卫浴间的黑白几何分割，均在古典工艺与现代美学之间，凝练收放自如的平衡美感。

When time goes still, and landscape is coagulated in traveler's view, the only patch of sea blue is overspread from the inner heart to complete the setting for the whole city of Suzhou.

If the places in the world are changing their places, then what would the world be like? E.g. Venice, a city honored Capital of Water, is moved into China. Just like Venice, the city of Suzhou has equivalent imagination of water. Sherwood Design incorporates images of blue sky and sea of Venice as well as mild Victorian style into this project, where occupants feel nothing but charm in a setting of dark and light hues and winding lines.

In the vestibule, the green of large area leads the visitors into a peaceful state of mind, which is like a lowering tone ready to come into its summit. In the hall, the bar area of western kitchen is coated in pure white and iron-grey mirror, embodying a strong dramatic tension. Three multi-arch curves are particularly used to expand the depth of field and offset the oppressing sensation by the low beam. The table, the chair and the lighting in the dining room are of white, light-grey and blue, where a retro fireplace is of delicate British style. When integrated with the TV wall in the living room, the fireplace accomplishes focus to avoid messy feelings by objects.

In order to continue the design theme, the area around the stairs is wrapped in dark veneer. At the stitching point, are LED lights, whose light and shadow set up an image of stars in sky, sometimes visible and sometimes invisible, the fun from which grounds off the stiffness and rigidness. Upward, the gold-paint handrail is rich in western flavor with solid wood flooring and crystal shaping the spatial magnificence. The master bedroom involves more colors of China blue-white porcelain and pale green to accomplish visual layers. The secondary bedroom is of floral wallpaper. The bed backboard features a totem done with rivets, generating feelings similar to the black-white geometric ratio in the bathroom and winding between classical and modern aesthetics to make a balance at a free wish.

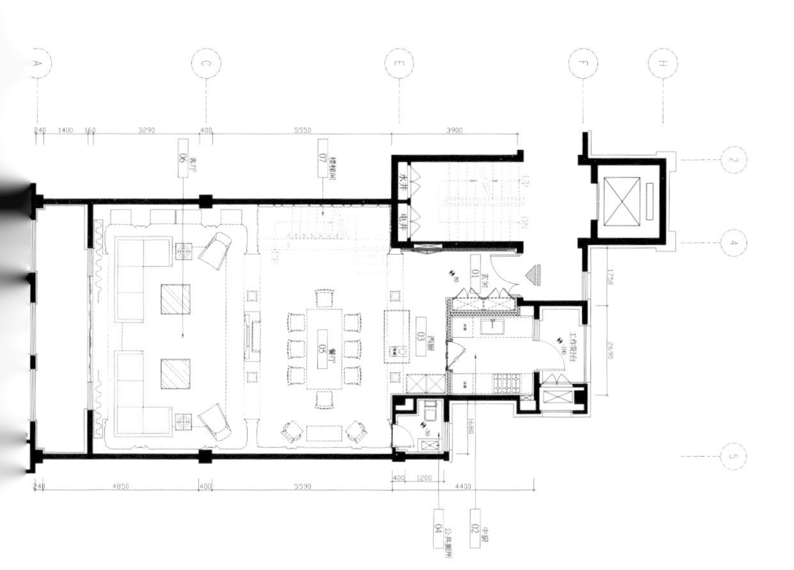

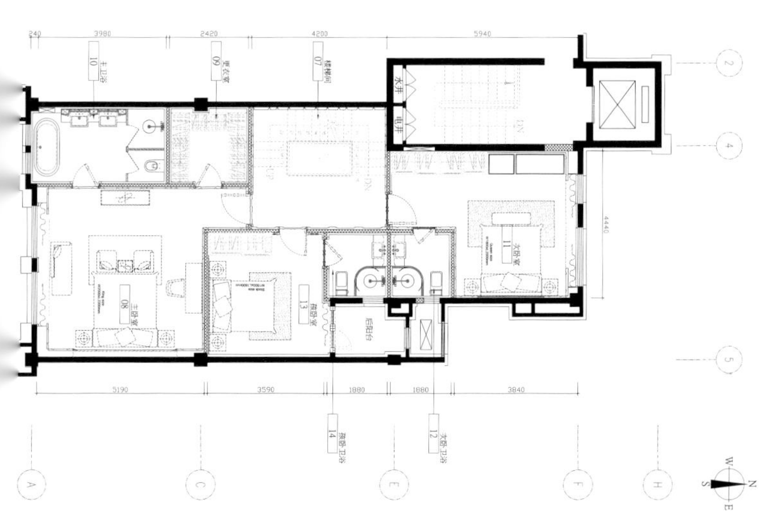

平面图 / Site Plan

The Untrammeled Soul Is from Spain

源于西班牙奔放不羁的灵魂

项目名称：好莱坞大公馆
设计公司：琳达建筑师事务所
摄影师：格雷·克劳福德、理查德·鲍尔斯

Project Name: Hollywood Renovation
Design Company: Linda Brettler Architect
Photographer: Grey Crawford, Richard Powers

不修缮，便无以展现本案空间最初的壮丽；不修缮，当今的生活方式便无以为继。鉴于空间的规模及现存的建筑特点，本案设计从色彩、软装、家具等方面着手，采取了一种“多便是多”的设计手法。

整个空间的作业始于外部风景。景观建筑师保罗与设计师琳达成功地把爬满绿藤的悬崖改造成一个室外的草坪，其间阡陌纵横。

白色灰泥的立面极富纪念意义，间以藤架、廊柱作为点缀。屋顶覆盖以赤土色的瓦。亮光盈盈的通道欢迎着客人的同时，也昭示着建筑优雅的历史。

入口处的布局及主要空间保持着旧时的特点。抹上灰泥的天花板以及装饰后的墙面让整个空间有了改头换面之效。玄关左边是客厅，客厅的西边悬挂有古董镜，镜面闪烁着镏金的光。玄关右边是一个嵌入式酒吧，酒吧的四壁覆盖着红色的墙纸，古色古香。由此向里，便是厨房。厨房里，现代化的设备尽情满足着家居生活的需要。邻近的仆人房早已拆除，扩大后的厨房空间，不但有了早餐室，即使烹饪准备等活动也有了各自的空间。天窗的彩色玻璃由本案室内设计师自行设计。主浴空间的彩色玻璃墙板，使通往楼上的通道也沾上了光彩。沿着光线，客人的脚步不由自主地通向了二楼空间。曾经拥有6个旮旯的漆黑卧室如今已华丽地转换成一个主卧及可供三个孩子使用的单独卧房。

主卧的空间，门厅真可谓是焕然一新。为了获得更多的自然光线，天花不仅得以抬高，窗户也得以增加。墙面柔软的蓝色涂料给人一种宁静的归隐之感，原本生动活泼的家具也因此有了更好的衬托。摩洛哥式的家具在活跃气氛的同时，更给人一种快乐、遁世的隐私感。

The house, was in need of a massive renovation to bring the property back to its original grandeur and to modernize it to fit today's lifestyle. Given the large scale of this home and the strong existing architectural features, Brettler embraced a "more is more" design approach in regard to layering ornate trimmings, fixtures and bold colors.

The transformation began with the exterior landscape. Working with landscape architect Dea Paul, Brettler transformed the formerly sloped space that culminated with an ivy-covered cliff into a usable outdoor lawn and meandering terraced paths.

The exterior is defined by monumental white-stucco facade that was interspersed with pergolas and porticos and is topped by a terracotta-tiled roof. The now light-filled entryway dramatically greets visitors and begins to reveal that home's elegant history.

The entry-level plan and primary rooms remain with original features, but the spaces received a facelift through furnishings and surface treatments, including artisanal plasterwork ceiling moldings and wall treatments. To the left of the entry is the living room where the western was has been faced with antique mirror to subtly reflect gilded glow. To the right of the entry, the refurbished dining room leads past a whimsical new bar alcove that is completed by a red vintage wallpaper panel, and into the kitchen. The kitchen was modernized to fit the lifestyle needs of a large family by removing a formerly adjacent maid's room and dividing pantry to create ample prep-area, eat-in space, and a larger breakfast room. Upstairs, a new stained-glass skylight designed by Brettler and vintage stained-glass panel backing the master bath brighten the hallway to introduce visitors to the fully reconfigures upper floor. What had felt "like a dorm" with six cramped and dark bedrooms has become a comfortable respite for the family of six with a master suite and three children's bedrooms. In the master suite, the architect updated the space by using the past as a key, raising the ceiling and adding a tray detail, as well as adding windows for increased natural light. Soft blue paint on the walls creates a tranquil retreat while serving as an understated backdrop to more lively furnishings. Moroccan-style furnishings enliven the space, creating a fun, private outdoor escape.

Flowers Making a Romantic March

花语呢喃，奏响一曲缠绵浪漫的进行曲

项目名称：中星红庐别墅 74 号别墅（法式）
设计公司：鼎族设计
设计师：吴军宏
摄影：三像摄建筑摄影机构 张静
面积：946.8 ㎡

Project Name: No. 74 Red Villa (French)
Design Company: Prosperous Clan Adorn Design
Designer: Wu Junhong
Photography: Threeimages Zhang Jing
Area: 942.8 m²

当巴黎的浪漫吸引了一批又一批的朝拜者，当福斯湾的诗情画意成为梦想的栖息地，当经百年时光雕刻的原欧别墅拨动心弦，原欧别墅的居住真义，穿越百年浪漫、荣耀与生生不息的建筑传奇，在这里——中星红庐别墅一一成为现实。

本案是中星红庐 74 号别墅，法式风情的设计风格，突出了居住空间的恢弘气势与浪漫气息，让人一旦进入，便沉醉其中。客厅法式的廊柱、精美的雕花、变化丰富的卷草纹样、鎏金的曲线……仿佛一切都如此自然，却又好似在崇尚冲突之美。也许你会觉得眼花缭乱，但你又不得不承认自己确实惊叹于其细节上的繁复且精细的做工以及最终凝聚起来的浓郁的浪漫色彩，它散发着人文、古典的气息与舒适、优雅、安逸的内在气质。一套经典优雅的鎏金茶具，一台老式高贵的拨盘电话机，一扇鎏金的花卉屏风，把欧洲贵族高贵典雅的生活艺术描绘得淋漓尽致。客厅的墙面加入了实用性和展示性兼具的壁炉，壁炉上方的镜面看似略显简单，却起到了提高空间的作用，使周围环境相互融合。

餐厅秉持典型的法式风格搭配原则，餐桌和餐椅个性却不张扬，朴实的外形，温暖典雅的颜色，体现出生活的底蕴。餐椅表面略带花卉图案，配合扶手和椅腿的弧形曲度，显得优雅矜贵。帘卷西珠，窗纱轻柔，搭配上水晶吊灯、壁灯、瓶插花束，浪漫、温馨之感扑面而来。而私密空间也延续了这般浪漫的风情，优雅的姿态，让你尽享酣眠。

浪漫、绅士是人们赋予法国的标签。法国人会生活，不仅仅体现在他们的浪漫爱情和社交的绅士上，也体现在其居住的环境上。中星红庐 74 号别墅的法式风格设计，着实让人体验了一把法国的浪漫情调。

XO
Hennessy

When the pilgrims have flooded the romantic Paris, the dwelling in dream has found its carrier in the idyllic of the Firths of Forth, and the historical European villa has drawn the attention of the ordinary people with its romantics that has been though a long time to carry forward the romantics and glory, the desire to live in an original villa in Europe has been realized in this project.

The French design in this project sets off the grand and the romantic in a living space, which makes people lost in it immediately. French columns, delicate carving, grass-rolling emblazonry and gliding curve seem to be inbuilt in the nature, making a kind of aesthetics that's conspicuously contrasting. You can't but marvel at its detailed, complex process, which gathers together to make strong romance. It's cultural, classical and comfort with an inner temperament of leisure and calm. Items of a set of gliding European tea ware, an old drive-plate telephone set and a gliding floral screen highlight the artistic life of European nobility incisively and vividly. The fireplace in the living room is both practical and aesthetic, the mirror above which looks relatively simple but functional to promote the visual effect to a new high level and blend the surroundings.

The dining room carries forward the French style to make configuration. The table and chair is unique but low-key. They are simple and earthy while their color is warm and elegant to embody details of life. Surface of the chairs is decorated with floral pattern but not very, which matches the handrail and the arched leg to look graceful and elegant. The coordination of pearl curtain, the gauze, the crystal chandelier, the wall lamp and the vase oozes romantics and warmth. The private space continues the romantic style, whose refined posture compliments your good sleep.

"Romantic" and "gentle" are labels people often use when speaking highly of France. French people know how to enjoy life. That's not only embodied by their romantic love and gentle social life but by their dwelling setting. Such is French style. It indeed makes a space where to appreciate French romantics.

Eastern Venice, Whispering by Light Green

东方威尼斯，湖绿轻吟

项目名称：苏州水岸中式秀墅
设计公司：台北玄武设计
设计师：黄书恒、林胤汶
软装设计：吴嘉苓、张禾蒂、沈颖
摄影师：王基守
撰文：程歆淳
用材：海南黑洞石、希腊白大理石、蛇纹石、金箔、酸洗镜、镀钛黑

Project Name: Beautiful Villa, Water Shore (Chinese Style), Suzhou
Design Company: Sherwood Design
Designer: Huang Shuheng, Lin Yinwen
Upholstering: Wu Jialing, Zhang Hedi, Shen Ying
Photographer: Wang Jishou
Text: Cheng Xinchun
Materials: Marble, Gold Foil, Mirror, Titanize Black
Area: 357 m²

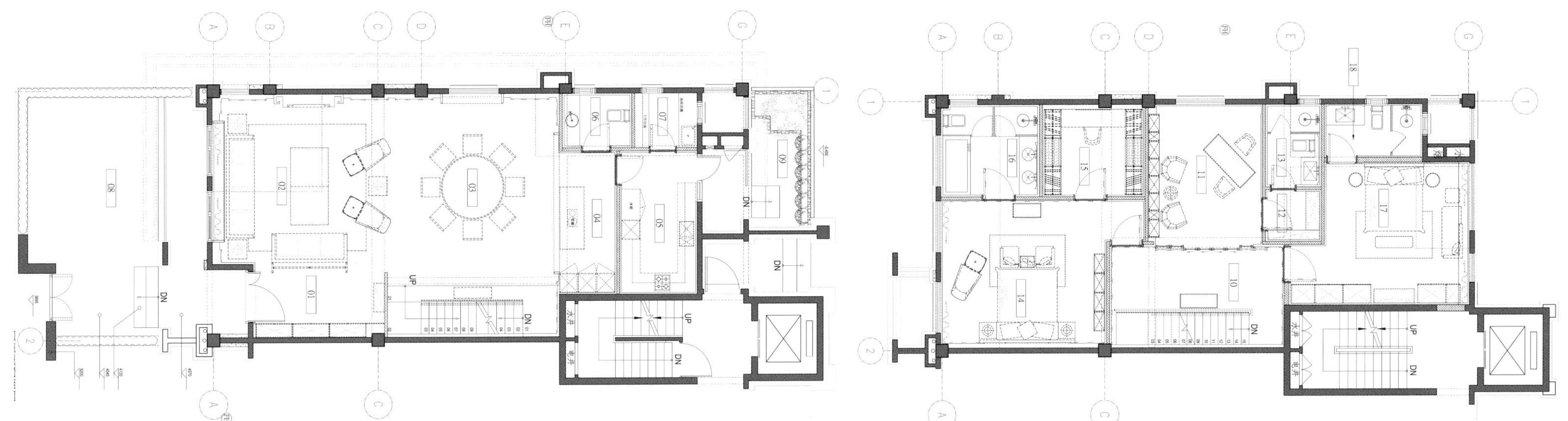

一层平面图 / First Floor Plan

二层平面图 / Second Floor Plan

苏州，一座水色盈溢的古老城市，与意大利威尼斯一样，具有绝佳的水乡风景与细腻的人文风情。春风拂面，细柳垂杨，清淡的城市笔触，总予人无限遐思，而建构于悠久历史上的现代景观，更使此地于中西交辩外，更呈现古今对话的可能，空间与时间的交错，铺就了苏州水岸秀墅的底蕴。玄武设计将《马可波罗东游记》作为故事主轴，以西方探险家与东方大汗的晤面机缘，巧妙转化为中西混搭风格，利用湖水色泽的深浅递变，于家饰的传统线条与硬装的现代材质之间，呈现专属于苏州的柔婉气韵。

踏入玄关，取材自知名建筑师莱特（Frank Lloyd Wright，1867—1959）的繁复窗花映入眼帘，装饰主义的流利线条，与对口鞋柜的金箔花样遥相呼应，体现东西元素的戏剧张力；几扇鎏金窗花深嵌壁面，为客厅增添古韵之余，亦成为串连视觉的利器，设计师进一步以镜面不锈钢天花的反射效果，调整空间比例，增强空间氛围，延伸线条起伏，餐厅以出风口串起内凹天花板，界定着客厅与餐厅界线的同时，亦使空间视觉更为开阔，彰显豪宅气势。

于色彩方面，玄武设计特以湖绿为底，将传统元素（如铜钱纹沙发）与现代工艺紧密结合，透过比例转换——如餐厅壁面的长条形，即是模拟竹筒质感，呈现古朴的东方韵味，二楼壁板虽为中式比例，侧面却以亮面材质藏匿花俏；或者色彩变奏——如客厅窗帘选用明黄跳色，转至卧室，便选以不同层次的草绿与黛绿等，于古意盎然的廊室内，体现“中西混搭”的风情——如马可波罗远渡重洋抵达中国，与忽必烈大汗把酒言欢、相互馈赠的和谐景致。

Suzhou, a city of water like Venice, enjoys good view and landscape that can generate infinite imagination with its culture, breeze in spring and willow in spring, summer and autumn. The modern landscape in a setting with a long history transfers Suzhou at the conjunction of the east and west while presenting a communication between the modern and the east, and between the space and the time. Such is a location where this project is. Beneath the skills of Sherwood Design, scenes of The Adventures of Marco Polo have found its stage in this space, where the meeting of the explorer from the west and the emperor of the east has been fused together, and the progressive change of lake hue, the tradition decoration and the modern material accomplish the morbidezza that's exclusive to Suzhou.

The vestibule features complicated floral windows initiated by Frank Lloyd Wright. Flowing lines of decorationism echo with the gold-foil pattern on the shoe cabinet, embodying a dramatic tension of the east and the west. Several gliding windows of lattice inlaid on the wall boast the charm and classical of the living room while stringing visual effect. The reflective effect of the mirror stainless steel ceiling shifts the spatial proportion, enhancing the spatial grander and generosity. The air outlet in the dining room joins the indent ceiling, partitioning the living room and the dining room and exaggerating the visual effect to outline the momentum of a mansion.

In a background dominated with lake green, traditional elements like the texture of copper cash on the sofa and those modern transferred in ratio, e.g. the bar modeling of the wall in the dining room, makes an imitation of bamboo and looks very primitive and eastern. Though the siding board on the second floor is very Chinese, the side is of glossy material and colors used complete a sharp but intended contrast, like the bright yellow of the curtain in the living room and the grass green and the dark green in the bedroom. The corridor is antique, a collection of the east and the west, where a feast scene of Marco Polo and Kubai Khan, the 5th emperor of the Yuan Dynasty, is available.

Mediterranean: Beautiful and Peaceful as the Hermit Kingdom

醇美地中海，遁世般的优美宁静

项目名称：波斯特住宅
设计师：理查德 · 兰德里

Project Name: Post House
Designer: Richard Landry

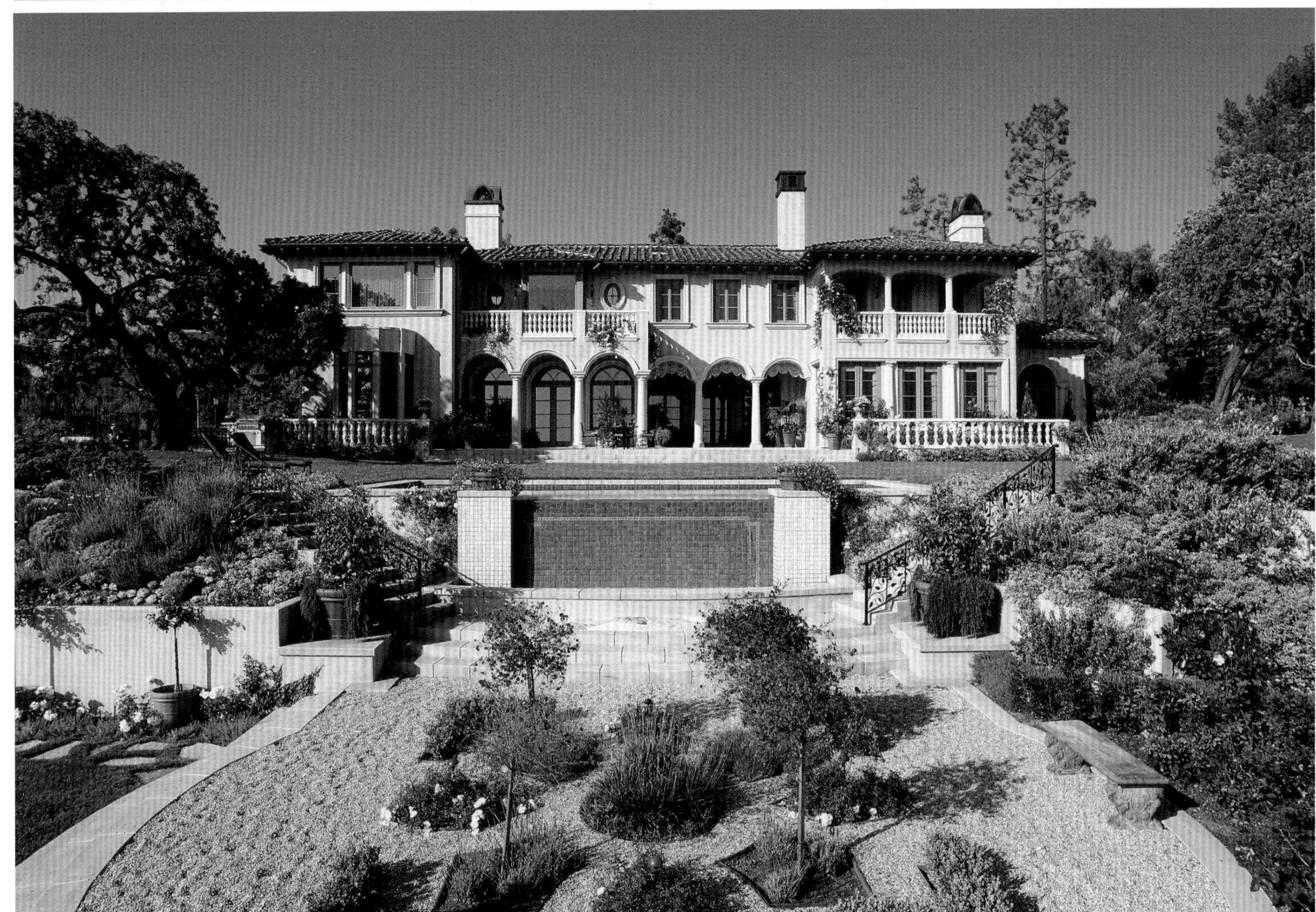

这栋约110平方米的房子拥有一个独一无二的空间环境，那就是坐落于小山顶上，在夜晚降临的时候可以俯览整个城市的灯光。它像是优雅的西班牙殖民时期风格的复兴，又似乎是一个伟大并且无可挑剔的好莱坞默片时代的遗存物。这里的别墅曾经由设计师华莱士·内夫为好莱坞明星葛洛丽亚·斯旺森和玛丽·皮克福德建造。虽然只是一幢房子，却似乎和探戈、查尔斯顿（一种舞蹈）产生了共鸣。现在，这座别墅是在几年前由兰德里设计集团在彻底重新建造的基础上设计的。

设计灵感来自于20世纪初华莱士·内夫和艾迪生·麦兹那古典的设计风格，以及那些坐落于意大利、西班牙和摩纳哥的山坡之上的地中海别墅。理查德·兰德里设计的这座新别墅，创造了一个新的时代经典。

为了提高设计的视野范围，兰德里和他的团队对西班牙、意大利和葡萄牙的房子做了广泛的设计研究。但结果绝对不是一个地中海别墅的复制，而是对地中海别墅的另一种阐释。兰德里设计团队使用柔和圆润的色调与光滑的灰泥来装饰墙壁，而屋顶使用的是罗马瓷砖瓦，栏杆和立柱的外观使用的则是预制的模拟天然石灰岩颜色的混凝土。

本案位于一个经典的意大利风格的花园中，其倾斜陡峭的地势给设计师带来了一些挑战。兰德里用阶梯式的花园解决了空间布局的困境，从而创建出更多梦幻般的房间。对这种古典的阶梯式地中海花园别墅，兰德里自己也极为称道。现在，这个向下倾斜的花园通向一个非常大的泳池，看起来如同奇迹般地悬浮于这个城市之中。

在室内，两层楼的大厅里是由理查德·兰德里专门设计的桃花心木和铸铁楼梯。楼梯优雅的向上旋转，最后深深地嵌入一个扇形的休息区，光线透过四瓣花窗洒满整个房间。

贯穿这个不规则伸展的房子，理查德·兰德里创建了自定义架构细节，包括雕刻壁炉地幔和空间，旨在展示业主精心收藏的艺术品和18世纪的古董。设计师小心翼翼地平衡着广阔的墙壁空间、绘画以及嵌入式壁龛，以期更好地展示雕塑及其他的工件。

这栋别墅最让人震惊的设计是图书馆。位于主轴颈的尽头，是脱离了其余房子的半独立屋，这是一个藏书家的私人密室。高高的天花板上，镶有一个眼睛形状的金箔，是镀金定做的樱桃镶板。图书室内设有一个极具特色的5.9米长的书架，可以使用一个可移动的梯子来阅读或者存取书架上的书本，而这里存储和展示的大量书籍，都是罕见和珍贵的第一版。

兰德里团队与艺术家兰斯·克莱姆为餐厅、盥洗室和楼上的墙壁创造出令人惊讶的视错壁画，再一次让人想起经典的地中海别墅，旧时代的典雅与印象，在这里便可尽享。通过令人敬畏的建筑细节比如立柱和高耸的穹棱拱顶的天花板，兰德里创造了一种遁世般优美宁静的意境。

Standing on a hilltop high above the city lights, the 110 square meters house has a remarkable aura of timelessness and permanence. A graceful Spanish Colonial Revival, it appears to be a grand and impeccably preserved relic of Hollywood's silent era, the kind of villa once built by Wallace Neff for stars like Gloria Swanson and Mary Pickford, a house that seems to echo with tangos and the Charleston. In reality, the house was built from the ground up by the Landry Design Group a few short years ago.

Inspired by the classic designs of master architects of the early 20th century Wallace Neff and Addison Mizner, as well as the Italian, Spanish and Monegasque villas which dot the hillsides above the Mediterranean Sea, Richard Landry designed a new house with timeless gravitas.

Landry and his team did extensive research on the designs of Spanish, Italian and Portuguese houses in order to distill their vision. The result is not a reproduction of a Mediterranean villa, but rather a re-interpretation. Using mellow-toned, hand-troweled smooth stucco and a Roman tile roof as their canvas, the Landry Design Group embellished the facade with balustrades and columns of pre-cast concrete in a limestone color that emulates natural stone.

The house is gem-set in a classic Italianate garden. The steeply sloping lot provided some challenge for the architect. Landry solved the spatial dilemma by terracing the gardens, thus creating the illusion of more room, and paying tribute to the stepped gardens of classic Mediterranean villas. Now, the down-sloping yard leads to an infinity pool that seems magically suspended above the city.

Inside, the two-storey foyer is dramatically framed by Richard Landry's custom-designed mahogany and wrought iron staircase, swirling upwards

towards a deeply inset scalloped recess framing a quatrefoil window that floods the room with natural light.

Throughout the sprawling house, Richard Landry created custom architectural details, including carved fireplace mantles and spaces intended to showcase the clients' extraordinary collection of art and 18th century antiques. He carefully balanced expansive wall spaces for paintings and inset niches designed to display sculpture and artifacts.

The library is one of the most stunning features of the Post house. Situated at the end of the main gallery, it is semi-detached from the rest of the house, a bibliophile's inner sanctum. The tall room, crowned with a gold leaf oculus in the ceiling, is clad with custom-designed cherry paneling. The room features 5.9-meter bookshelves-accessed by a movable ladder-to store and showcase the client's vast collection of rare first editions.

The Landry team worked with artist Lance Klemm to create surprising trompe l'oeil frescoes for the dining room, powder room and upstairs gallery. Again reminiscent of classic Mediterranean villas, they add another layer of Old World elegance, enhancing the impression of age. By using august architectural details like columns and soaring groin vault ceilings, Landry created an atmosphere that is almost cloistral in its polished serenity.

Magnificent Rebirth of Tranditional Architecture

传统建筑的华美新生

项目名称：意大利乡村别墅
设计公司：兰德里设计

Project Name: Villa Del Lago
Design Company: Landry Design Group

“意大利乡村别墅”占地2 317平方米，建于悬崖峭壁，乡野风情，给人一种质朴、欢迎的感觉。梁柱、地砖、屋瓦以可回收材质制作，空间因此有了一种久远的时间感。隐私、半隐私、开放的空间无缝衔接于一体。宏伟的双层玄关，具有恢宏的气势。屋面凉廊开以双折叠门楣。温暖的季节，双面洞开，自是一片暖阳。

双层的墙体，以其厚实对比着旋转楼梯的精致壁板与性感曲线。多媒体室的玻璃地板，惹弄着室内人的眼。无意中一瞥，是石头墙、木梁柱佑护下的“爱车”。室内各空间全开向庭院，远观除了碧绿的湖还有皑皑群山。

除此之外，“定制”的空间还糅合了一些环保元素，如再次循环利用的供水系统、污水回收处理体系、LED照明、供热制冷等等。量体之上，石缝之间，镶嵌着太阳能板，反射进公用设施的太阳能源是低碳，也是环保。

Built in the style of a rustic Italian villa, the 2,317-square-meters bluff-side residence belies its magnificent size with an earthy, welcoming feel. Villa del Lago appears to be a home that has grown over time. Reclaimed beams, tile and roof tile add an instant patina of history and a hint of Old World charm into the residence. Inside, Landry orchestrates a medley of private, semi-private and public spaces that seamlessly flow together. A grand entrance hall gives way to a two-storey great room. Covered loggias featuring bi-fold doors open up to the skies during warm weather.

A two-storey stone wall's bulk and solidity contrast with the refined wood paneling and the sexy curves of a circular staircase. Cool glass adorns the floor of the media room, giving visitors a tantalizing glimpse of sleek cars sheltered within the stone walls and wood beams of the garage below. While offering sturdy shelter, each room in the home opens to a resplendent yard, leading the eye outward toward the blue lake and the mountains beyond.

Furthermore, this custom estate incorporates several environmentally friendly elements. Solar panels are concealed on the hill above the house and are projected to result in no electric bill from the public utility. An on-site water well, re-circulated water systems, grey water reclamation, LED light fixtures throughout, and high efficiency heating and cooling systems are among the other environmentally friendly elements. These features are concealed by locally cut stone for the facade, reclaimed antique roof tile and wood beams from Italy, and stone floors that were collected from an abandoned chateau in France.

CHAMPION

染上宝石蓝的美式新风尚

New American Fashion Dyed with Jewelry Blue

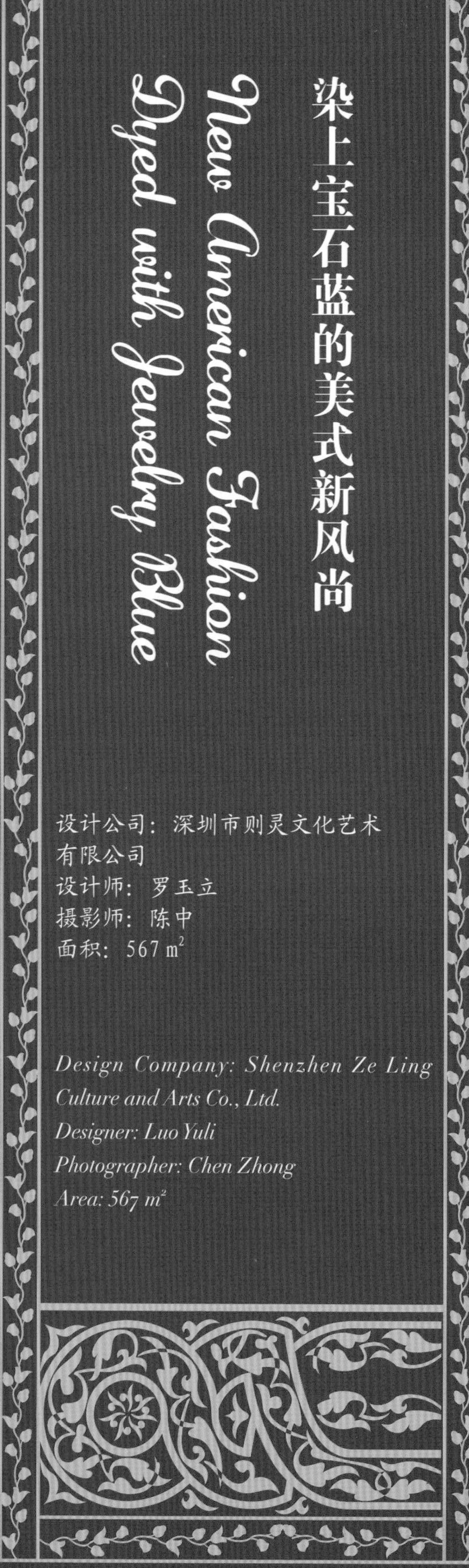

设计公司：深圳市则灵文化艺术有限公司
设计师：罗玉立
摄影师：陈中
面积：567 m²

Design Company: Shenzhen Ze Ling Culture and Arts Co., Ltd.
Designer: Luo Yuli
Photographer: Chen Zhong
Area: 567 m²

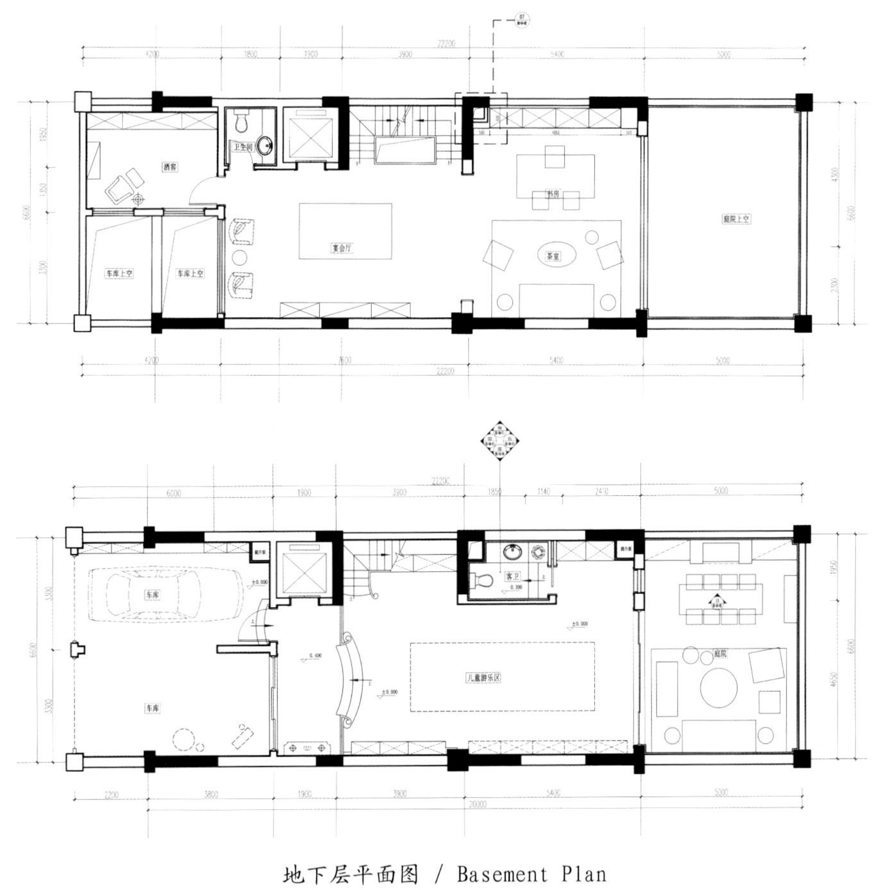

地下层平面图 / Basement Plan

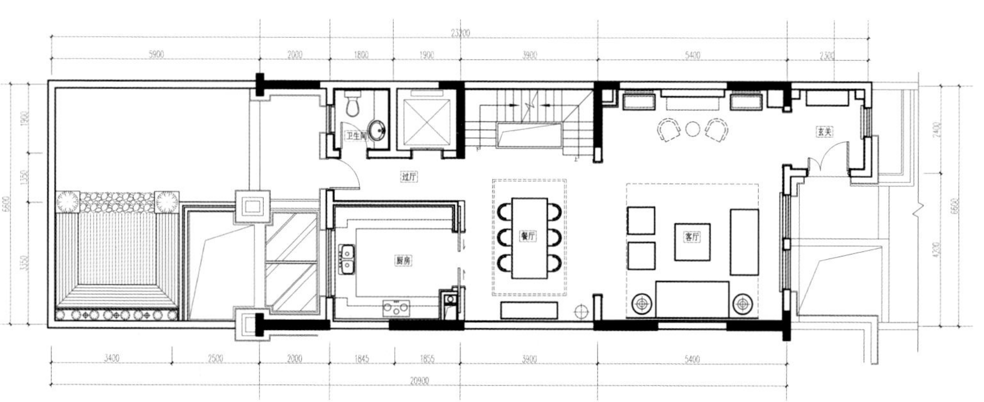

一层平面图 / First Floor Plan

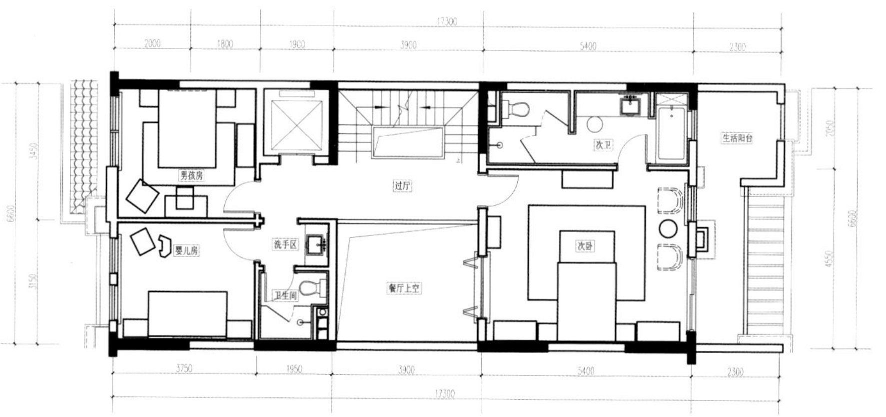

二层平面图 / Second Floor Plan

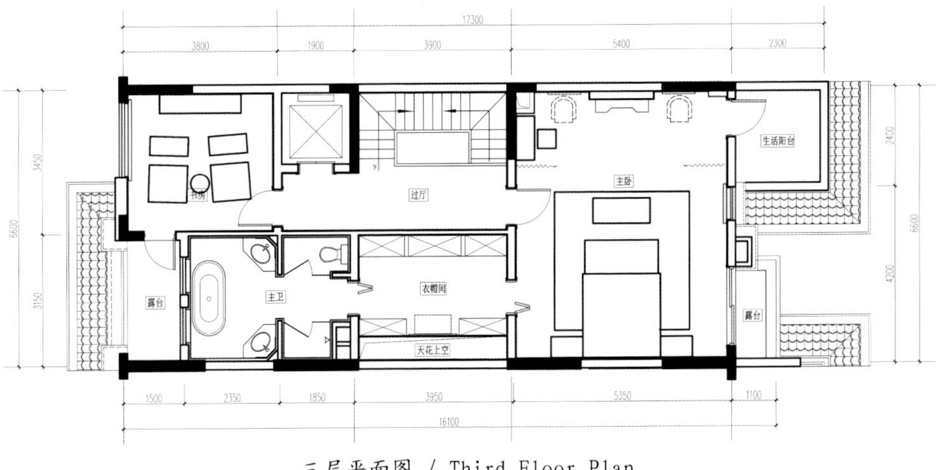

三层平面图 / Third Floor Plan

美式风格受到了美国文化的深刻影响，追求自由的美国人把舒适当作营造居住环境的主要目标。美式家居浪漫自由的生活氛围，让都市人消除了工作的疲惫，忘却了都市的喧闹，拥有了健康的生活与浪漫的人生。这正是我们这个高速发展的时代越来越多的客户渴望的生活方式。

室内色彩的规划上以蓝色调为基础，在墙面与家具以及陈设品的色彩选择上，多以自然、怀旧、散发着质朴艺术气息的色彩为主。整体朴实、清新素雅、贴近大自然。山水图案的床品搭配柔软布料，使室内充满了自然和艺术的气息。从窗外洒落进来的明媚的阳光，在富有生命力的绿植的点缀下，给整个空间带来愉悦、充满活力的生活氛围。让身处其间的主人，感到由衷的舒适，满怀生活的愉悦。

平面布局整体大方，轻松优雅，体现出美式风格舒适、不拘小节的特点。功能分区明确，将居住功能与社交功能适度隔离，既保障主人在居住空间里有良好的私密感受，又重点强调出别墅空间不同于一般公寓空间的社交与娱乐功能，让客户自由享受高端生活的美好。

强调面料的质地，运用手绘的大自然图案的墙纸、斗橱、布艺等饰品为居室营造出独特的自然气息，符合现代人的生活方式和习惯，再加上绿植等自然景物的搭配，使居住的人感受到轻松、舒适的身心享受和居住体验。以凸显主人追求简约、自然环保的新时代的价值观与人生观。

American style is exerted with a great influence of American culture, where American seeking freedom takes comfort as their main goal when it comes to residence. The life air of dwelling freedom and romantics in America keeps the fatigue away from the urban people to have a healthy and romantic life. And this is what more and more people occupied in city have increasingly been desiring.

The interior is dominated with blue, while walls, furnishings and accessories focus on a natural, nostalgic, and primitive sense. Close to nature, the holistic is simple, fresh and elegant. The bedding patterned with landscape instills the space a natural but artistic atmosphere. The sunlight through windows set off with the vigor and vitality of green plants offers pleasure, happiness and vitality. People there feel nothing but comfort in a good mood to live. The whole layout is in good taste, relaxing and elegant to embody features of the American style comfortable but not sticking to trifles. The functions are clearly divided into living and socializing, ensuring a good privacy for the family while a villa is quite different from the usual ones in terms of social life and entertainment, so that a high-end life can be guaranteed.

The textured fabric, the wall with the nature hand-sketched, the chest of drawers and the fabrics set up a unique living atmosphere. This is in good line with the modern living style and habit. The green plants allows for mental enjoyment of relaxation and comfort to highlight a personal value and life attitude to pursue simplicity and environment protection.

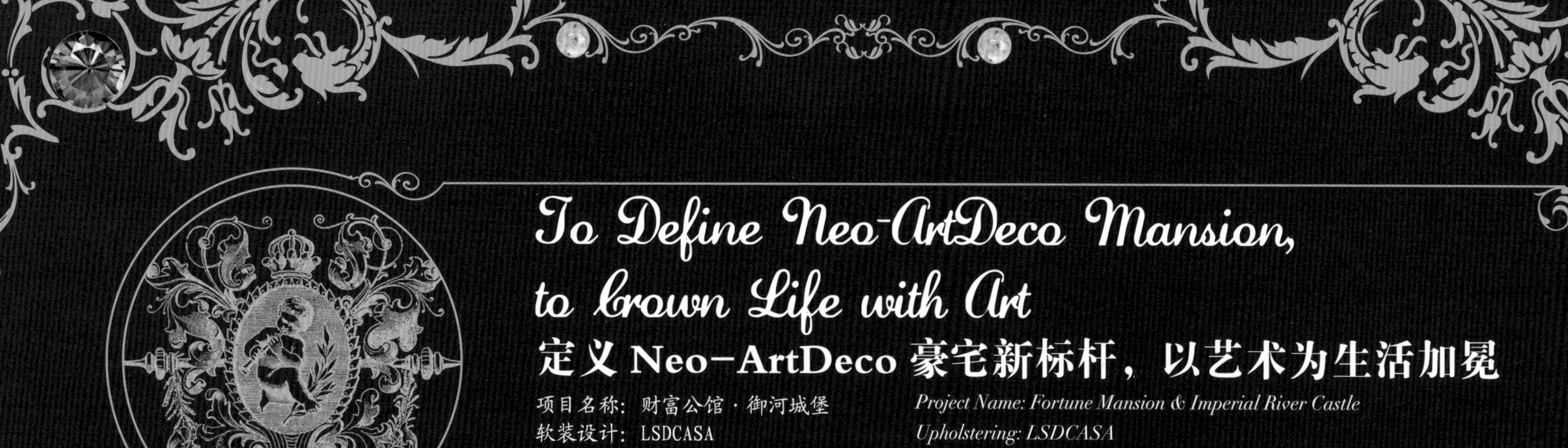

To Define Neo-ArtDeco Mansion, to Crown Life with Art

定义 Neo-ArtDeco 豪宅新标杆，以艺术为生活加冕

项目名称：财富公馆·御河城堡
软装设计：LSDCASA
设计师：葛亚曦、蒋文蔚
地点：中国北京
面积：1 600 m^2

Project Name: Fortune Mansion & Imperial River Castle
Upholstering: LSDCASA
Designer: Ge Yaxi, Jiang Wenwei
Location: Beijing, China
Area: 1,600 m^2

中国十大别墅之一——财富公馆·御河城堡地处北京朝阳区，是财富地产集团为名门贵胄打造的领航产品，被誉为中国顶级别墅代表。

该别墅样板间五年前由知名设计师邱德光先生设计，因其大胆张扬的风格在当时被认为是邱先生新装饰主义风格的代表作。中国设计思潮发展非常迅速，人们对设计的要求不再停留在某设计风格的样式表层，从单一的追随转向深层的思考和面对自身的独特需要，注重挖掘能对应精神认同层面的需求和设计本身应有的人文关怀。财富集团为了再造中国顶级新装饰主义豪宅，特别委托 LSDCASA 担纲财富公馆·御河城堡的别墅样板间改造，期待展现设计特点的同时，再次代表财富阶层的生活。

LSDCASA，延续了建筑及室内的新装饰主义风格，续写着丰沛的美学力量空间，以匹配财富阶层应有的生活方式。当今世界，科技及生活的发展为人们提供了各种程序和解决方案，无谓追求单一的设计审美，让多元化争论和质疑互为存在，设计在本案中再次发挥创新的力量，打破既有程式，让单一的权力、财富的显性诉求，过渡到生活中对伦理、礼序、欢愉、温暖的需要。

这套 1 600 平方米的府邸共有三层，从地下一层逐步向上，空间的每一层都有自己独特功能和对应的趣味和隐喻。

门厅，室内设计的调整配合建筑空间倾向于表现特质与规律的设计意图，故门厅处保留了近似公共空间的尺度，给人的压迫感的力量，把墙面常有的明显后现代徽章图案的墙纸更换为层次稳定、有力量感的色彩，配合意大利基于传统审美却又藐视规则、大众习俗的设计和装饰艺术，让空间拥有近似庙宇或会议厅般的神秘庄重。

客厅贯穿门厅的设计风格，设计师在家具陈列上采用了强烈的对称和仪式感，色彩是这里最好的礼赞，以冷艳高贵的钻石蓝与沉稳大气的咖啡色为色彩基调，搭配璀璨的金色和经典的黑白色，从天花到四周，从家具到靠垫，从饰品到绿植，无不展现待客空间的华贵。

西餐厅以沉着的墨绿色为主色调，搭配浅绿的窗帘帷幔，点缀蓝色与白色的精致花艺，餐厅一隅，雪白的孔雀拖着一袭长尾妆点着华美的空间，让这个中西融合的空间层次起伏，生机盎然、鸟语花香。风格独具的中餐厅则兼融了大户宴客排场和文人精神，餐厅古典实木家具，精致黑白插画的屏风，辅以餐厅中精美的花艺点缀，糅合出平衡典雅的用餐氛围。

走上二层，可以看到一个个色彩平衡、层次丰富的卧室空间。米色和咖啡色系是这里最经典的色彩基调。在这个基调上，设计师利用不同层次的紫色和蓝色，一时轻快、一时沉稳，为不同的主人营造韵味十足的私密空间。

书房，是主人收藏与展示的空间，各种藏品

和摆件演绎出空间格调，通过从细节到整体的微妙处理，男主人温文尔雅的外表之下，对品质生活的追求得到了完美的诠释。

地下一层作为主人娱乐和休闲的区域，分布着恒温泳池和影音室，纵观整套府邸，更像是具备魅力和非凡感官的艺术臻品，时光就此凝练成艺术，生活由此完美升华。

This project, as one of Top 10 villas, is a leading project by its owner and honored as the high-end representative among the same level in China.
Five years ago, the sample house of the real estate was done by Qiu Deguang, a famous designer, whose bold style was thought a magnum opus of neo art deco by Mr. Qiu. In the past years, design in China has undertaken a rapid development, where design requirements have shifted from the single follow-up to personal demand instead of the skin layer of a style by stressing the spiritual respondence and the humanistic care inborn with the design. And now, in order to present a top mansion of new art deco in China, the refurbishment of the sample house ever by Qiu is aimed to bring out features of design and life of the wealthy elites under the hands of LSDCASA.
On the basis of continuing the art deco and the spatial aesthetics, LSDCASA makes innovative efforts to break away the stereotyped practice and transmit the monotonous resort of power and wealth into needs of ethic, ritual, happiness and warmth. All are met for matching the lifestyle of the rich group, not seeking dull design for people with multi choices in a modern setting of science and technology, and taking plural arguments and questions under the same roof.
Of the 3 floors of 1,600 square meters, each is endowed with its own function, interest and metaphors.
In the foyer, the interior coordinative to the building is intended for presentation of a unique style. The tension similar to that by public space is kept; the post-modern wallpaper is replaced with hue layers that's stead and powerful. Design and art deco in line with aesthetic tradition but defying rules allow the space with a mysterious solemnness alike to that of temple and conference hall.
The same style is carried out throughout the living room, where furnishings and accessories are rich in symmetry and ritual sense. Color makes the lightest spot. The tone of indigo and brown is interspersed with gold, black and white. From

the ceiling to the horizontal setting, all items like furniture, cushion, ornament and greenery confides in nothing but luxury.

As for the western dining room, its hue is blackish green, which is accompanied with curtains and mantles of light green, as well as delicate flowers of blue and white. In one corner, a snow-white peacock makes a pose with its long tail. The Chinese dining room is inclusive of requirements of rich and influential families to hold banquet and literati's spirit. The classical furniture of solid wood and the black-white inset screen with floriculture complete a balanced and elegant atmosphere.

The second floor is world of balanced colors and rich layers. Beige and brown are the most classical, into which are fused purple and blue of different layers to build up private space with a lasting appeal, though sometimes light and then staid.

The study is for personal collection and display, where articles from the detailed to the holistic make a perfect interpretation of the quality life the button-down host pursues.

The underground is used for entertainment and leisure to accommodate swimming pool of constant temperature and audio-vision room. The whole space is rather like an art piece with special charm and extraordinary organ impact, where time is developed into art and life is then sublimed perfectly.

Fusion Classical and Contemporary: Conservative, Luxurious and Connotative

古典与时尚结合，低调奢华有内涵

项目名称：中星红庐 71# 别墅
设计公司：鼎族设计
摄影：三像摄建筑摄影机构 张静
面积：527.3 m²

Project Name: No. 71, Red Star Villa
Design Company: Prosperous Clan Adorn Design
Photography: Threeimages Zhang Jing
Area: 527.3 m²

这套住宅的整体空间开放流动且层次丰富，各个区域集中紧凑又相互渗透。在一楼的布局上，设计巧妙地挤出了中西两个厨房和早餐区，并和餐厅起居室相连，形成了一个完整的供家人就餐、休闲的场所。起居室旁布置了一间老人房套间，方便老人的生活起居。二楼充分利用空间，缩短了走廊的面积，安排了主卧套房、次卧套房和一个起居室，空间紧凑实用。地下室空间虽小，但也根据主人的个性爱好，量身打造了视听室、桌球区、酒吧区、健身房和棋牌室，格调高雅，样样俱全，并设置了佣人房和佣人卫生间、洗衣房。此外，设计还对楼梯进行了改造，从而把佣人房的房门开在楼梯间之下，尽量避免影响主人的生活起居。

在风格的设定上，设计采用了古典新奢华主义风格，把欧式古典风格和现代时尚元素相结合，纯正的古典装饰元素和新奢华主义风格的家具、陈设相碰撞，产生了极佳的视觉效果。作为欧式古典风格的常见元素，设计师以精致的帷幔作为室内的隔断，赋予了空间梦幻般的气质。

客厅坡顶以重复排列的几何图形装饰，并搭配镜面，延伸空间视觉。墙上油画的色彩也与水晶吊灯和室内整体色调形成了完美呼应。餐厅与会客区里，古典家具一改常见的脂粉气息，而以深紫色系为主，更加突显银色外廓的优美与华丽。厨房采用开放式设计，与早餐厅相连，结合欧式古典风格常用的镜面元素，大大拓展了空间视觉，通过反射形成的透视效果也令人感觉通透舒适。二楼小客厅的白色护墙板营造出亲切、温馨的气氛。淡雅的色调和笔直的线条打造出简洁而富有理性的空间。地下室一对拱形壁龛围绕着真火壁炉，让品酒区散发着令人沉醉的浪漫气息。

卧室以深色调彰显出冷艳与奢华的气质，古典家具的线条与卷草纹样给室内带来了典雅浪漫的气息。

总的来说，本案保留了古典的温馨、浪漫和文化品位，又被赋予了强烈的时代气息，处处散发着贵族的气息和低调的奢华。

Open and flowing on the whole, the project is of multilayers with all sections compact and interspersed. Kitchens eastern and western and an area for breakfast attached to the dining room and the living room on the first floor makes a field to dine and have leisure in a whole sense. Beside the living room is a suite for the elder, easy and convenient for them to enjoy life. On the second floor, the corridor is shortened for the accommodation of the master and the secondary suites. Though small, besides the maid room and the bathroom for the maid, the basement is fixed with audio-visual room, snooker room, bar, gym and chess and card room. All are in line with personal taste of the owner. The transformation of the stairs contributes to the door to the maid room which is positioned under the stairs, so life of the owners can be free of any disturbance.

The employment of the neo-classical luxury combines the classical European style and the modern elements, where authentic classical decorative elements and neo-luxury furnishings and accesses exert a terrific visual effect in collision. Streamers, a usual element in classical European style, now serve as the spatial partition to allow for a dreamy temperament.

The sloping ceiling of the dining room is decorated with repetitive geometric pattern, which with mirror extend the visual effect. The color of the oil painting on the wall and the chandelier echo with the internal hues perfectly. In the dining room and the reception area, the classical furnishings focus on dark purple instead of the usual color to make the silver external contour more beautiful and magnificent. The kitchen is designed open and connected with the breakfast hall, where the mirror element widely-used in classical European style expands the spatial vision to a large degree. The transparency effect by reflection enhances the spatial comfort. The white chair rail in the small living room on the 2nd floor builds up a warm, amiable ambience. The elegant colors and the direct lines shape a space concise and reasonable. The pair of arched niche oozes a romantic sense around the wine-tasting area.

As for the bedroom, dark hues set off the quiet and magnificent luxury. Along the lines and grass-rolling pattern of classical furnishings, comes a romantic sense.

On the whole, the project keeps the classical warmth, romantics and cultural taste while endowed with a strong modern sense, so everywhere are the noble sense and the conservative luxury.

VARVADA

东西兼容的独特魅力：当江南园林遇见新装饰主义

Compatible of the East and the West: When Southern Gardens meet New Deco

项目名称：杭州桃花源西锦园
设计师：邱德光

Project Name: Hangzhou Utopia
Designer: Qiu Deguang

中国人在21世纪应该过什么样的生活？住在什么样的房子里？多年来新装饰主义大师邱德光一直在思考这个问题。

在他主持的杭州桃花源设计案中，有了突破性的解答——在中式园林的框架下，让中国情怀与现代巴洛克、ArtDeco进行对接，也让其新装饰主义风格的成就达到前所未有的高度。

邱德光认为，现代中国人不能回到过去，过着古人的生活，既不便利也不实际；也不能把西方现代主义直接搬过来，过着西洋人的生活，而完全丧失中国固有的精神与文化。

在杭州桃花源西锦园一案中，在仿古的中国园林的环境背景下，他没有把中国古典家具搬进来，也没有原汁原味地采用中国传统语汇，而是选择与现代生活对接，对各种风格进行筛选，最终采用新装饰主义语汇。因项目位于杭州，人文气息浓郁，设计便把中国文化历史艺术与西方现代巴洛克、ArtDeco混搭，创造了一个既具有民族个性又兼容现代功能的空间。它是属于世界的，以东方风格为灵感，并很好地诠释了现代国际生活。

走进杭州桃花源西锦园，外面回廊蜿蜒、小桥流水、山水洞天、亭台楼阁，是中国园林精华的再现，是第一层的桃花源；里面又以西方现代主义的硬件形制对照精雕细琢的中国简约图腾，西式壁炉对照明式圈椅，东西混搭出另一层的桃花源。

该案设定屋主为具有国际观的收藏家，他既收藏了中式园林，也收藏了中西合璧的完美现代生活。在邱德光的主持下，这里潜藏了一个当代中国人的现代桃花源，不仅传承了中国传统的园林精神，更将其转换成一种新时尚，实实在在地照进生活。

厅堂具世家情怀又彰显中国文人意境，以金箔打造的巴洛克穹顶、尊贵的大理石柱、黑色稳重的水晶灯营造出大户人家的气势。中国式图腾衍生的绣花地毯、中国雕花般细腻的家饰陈设，则将满腔的中国情怀融入其中。为了让中国古典风格不显厚重，也为了衬托出屋主的年轻飞扬气质，设计在家具陈设方面采取混搭手法，既有明式家具，也有B&B、Cassina等西方极简风格代表家具，还有中国风的Baker家具，以及中西融合的创新家具，让东西方风格完美交汇。

餐厅与厨房为开放式空间布局，是现代机能与东方灵感的绝佳组合。伴随舒适便利的西式吧台、现代化厨具、巴洛克风格家具，中国山水画中善用的墨绿铺陈其中，墙壁壁面以繁复细致的中国式图纹雕花，创造东西交融的雅致和谐局面。

起居室、书房、卧房以及玄关走廊，都是东方与西方文化交遇的温柔典雅，游走于古典气息与现代时尚之间，例如中国景泰蓝、瓷器、案几等，与共同形成东西兼容并蓄的独特魅力与气质，让具有国际视野与胸襟的现代中国人得以抒怀……

What kind of life should Chinese have in the 21st century be? What kind of house should they live in? These are questions new Deco master Mr. Qiu Deguang has been thinking about over the years. The project "Hangzhou Utopia" by him offers a breakthrough solution. That is, in a setting of Chinese garden, the Chinese spirit and modern Baroque and ArtDeco are combined, which in turn promotes his new decorative style to a new level unprecedentedly.

According to Qiu Deguang, the idea, that modern Chinese people return to the past to have an ancient life, is neither convenient nor practical, while a completely western life, if used here, actually would lose the spirit and culture inherent in China's culture.

In the project against surroundings with antique Chinese gardens, Mr. Qiu does not move into the Chinese classical furniture, nor does he employ authentically traditional Chinese vocabulary. Instead he starts by combing modern life and selecting from a variety of styles to use the new Deco elements. Because the project is located in Hangzhou, a city with a rich literary atmosphere heavy, he mixes up the Chinese cultural history, the modern Baroque, and Art Dec, creating a nation with national personality, but compatible with modern function. It belongs to the world, but goes back to the oriental inspiration, completing a style that makes modern interpretations of international life.

The exterior is a Chinese garden reproduction with winding corridors, bridges, flowing waters, hills, caves, and pavilions, while the interior makes a world of mashup, where the inherent spirit of Chinese simple totem is fused with modern western furniture and accessories like western-style fireplace that contrasts with the eastern arm chair.

The owner is supposed to be a collector with a worldview, who not only collects Chinese garden, but also modern life with an international outlook. Under the direction of Qiu Deguang, here makes a modern paradise for contemporary Chinese people, which not only pass down the spirit of the traditional Chinese garden, but converts it into a new fashion of daily life.

The hall with Chinese literati family feelings and mood where to have a baroque gilded dome, noble marble columns, and black crystal lights, provides a kind of momentum exclusive to well-renowned large family. Chinese-style embroidered totem carpet and Chinese furnishings as delicate as carving implant

the space with Chinese feelings. To soften the classical Chinese style and set off the vigor and vitality of the young owner, furniture of the east and the west are mixed up: when some are of Ming style, there are representative items of Western minimalist style furniture, like B & B, and Cassina, as well as Chinoiserie Baker and innovative furnishings.

The dining room and the kitchen are designed with combination of modern function and oriental inspiration. Western-style bar, modern kitchen ware, and Baroque-style furniture are embellished with Chinese landscape painting. Walls are fixed with delicate Chinese graphic carving. All together makes a harmony of the east and west that's elegant and harmonious.

The entrance hallway, the living room, and the bedroom continue the refined and the mild by fusing the cultural exchanges between East and West. From the blend of Chinese cloisonne, porcelain, long table, western Baroque chair, painting in their abstract sense, and crystal lamps, comes a kind of inclusive charm and nature,. Consequently, modern Chinese with global vision can pour out their heart here.

Indigo Aesthetics of Customized Home

醉紫淀蓝，定制家的个性之美

项目名称：豹别墅
设计师：Henrique Steyer

Project Name: Leopard
Designer: Henrique Steyer

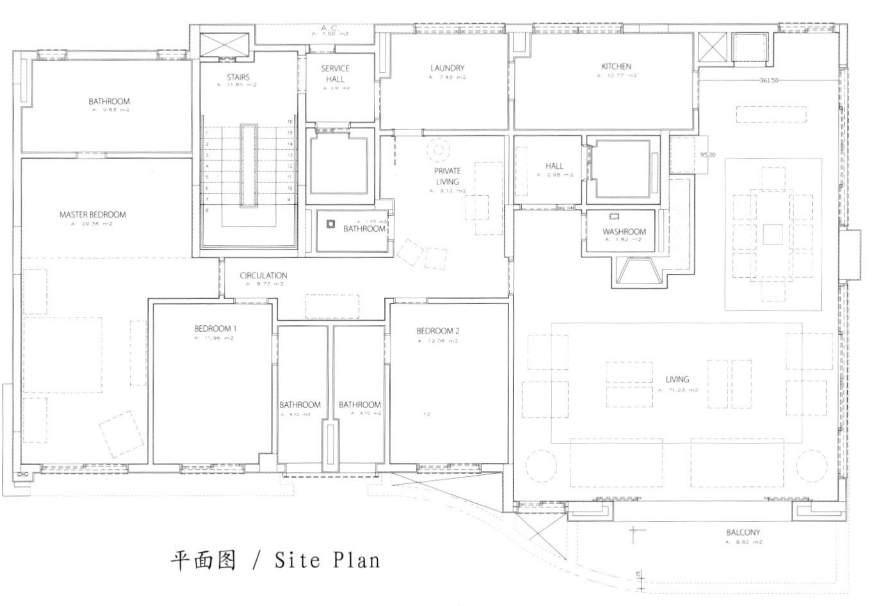

平面图 / Site Plan

国际顶尖的家具、法式古董使地处巴西南部的“豹”别墅更加奢华。与众不同的元素，如紫色的窗帘、富有视觉冲击力的艺术品，以其世界性的精髓，把本案转化成一个凝聚的统一体。意大利的顶级品牌沙发、卧榻与经典的品牌合二为一。法式的古董家具对比着现代性的空间语言。设计将20世纪70年代情色电影里的形象与令人玩味的主题很好地植入空间中，吸引着众人的注意。地板上铺着大块的地毯，升华着希腊大理石的雕像与金色的入户门。

客厅里，意大利式的沙发边配有两个台灯，并以“龙”雕塑为伴。台灯、雕塑全是大师手笔。餐厅里，亚克力材质的移动式桌子配有意大利式的金色圆墩，与烧烤架形成了极为相衬的配套，餐厅的灯光也均为大师手笔。大厅的金色大门旁，有气势逼人的法式镜，镜子的旁边就是电梯。

私人生活区，有一个亮光的组柜，如同客厅里的家庭影院。柜子上，除了一个小小的酒箱，还陈列有一些艺术品，其中一个便是如同水晶般的豹子。阅览区有一张英式古董书桌。卧室区，有一张桌子和一个窄窄的梳妆台，两件都是法式古董，正好充当了床头柜，白色的灯组也为卧室披上了亮眼的光辉。

浴室里，订制的木件给人一种手工的质感。洗手间里，一排古董镜自地板到天花沿墙面依次而立。厨房里，深紫色的组柜搭配着水泥色的可丽耐顶。

Top international furniture a
with a series of French antique p
confers luxury and irreveren
this apartment in southern Br
Unusual elements, like pink cur
and impacting artworks make
project with cosmopolitan essence
a very singular one. Top sofas
couches from the Italian Flex
combine with classic items from br
like Poliform and Flos. Contra
with the contemporary languag
the project, a series of antique Fr
furniture completes the decor. Am
the artworks, attention is drawn t
one by Mark Gary Adams, a ficti
artist created by Henrique itself,
images from 1970s porn movies m
with playful themes. On the floor,
rugs spread thru the rooms, which
boasts a leopard skin, a marble G
sculpture and a gold-plated entr
door.

On the living room, Italian sofas s
space with a pair of table lamp
Achille Castiglioni for Flos, and
a dragon sculpture by Lalique. By
dining room, an acrylic side table
Italian golden stools serve as suppor
the barbecue pit. The highlight of
room is the Zeppelin Lamp, designe
Marcel Wanders for Flos. In the h
the gold-plated door is accompanie
an imposing French mirror next to
elevator.

On the private living room, a high-g
cabinet, like the one of the living r
Home-Theater, holds a small wine h
and art objects like a crystal Bacc
panther. An antique English desk is
reading area. In the bedroom, a t
and a narrow dresser, both Fre
antiques, serve as nightstands, ligh
by white Flos lamps.

In the bathroom, custom woodw
gives a handmade feel to the proj
In the washroom, a great collection
antique mirrors line the wall from fl
to ceiling. In the kitchen, eggpla
colored cabinets match the ceme
colored Corian kitchen-top.

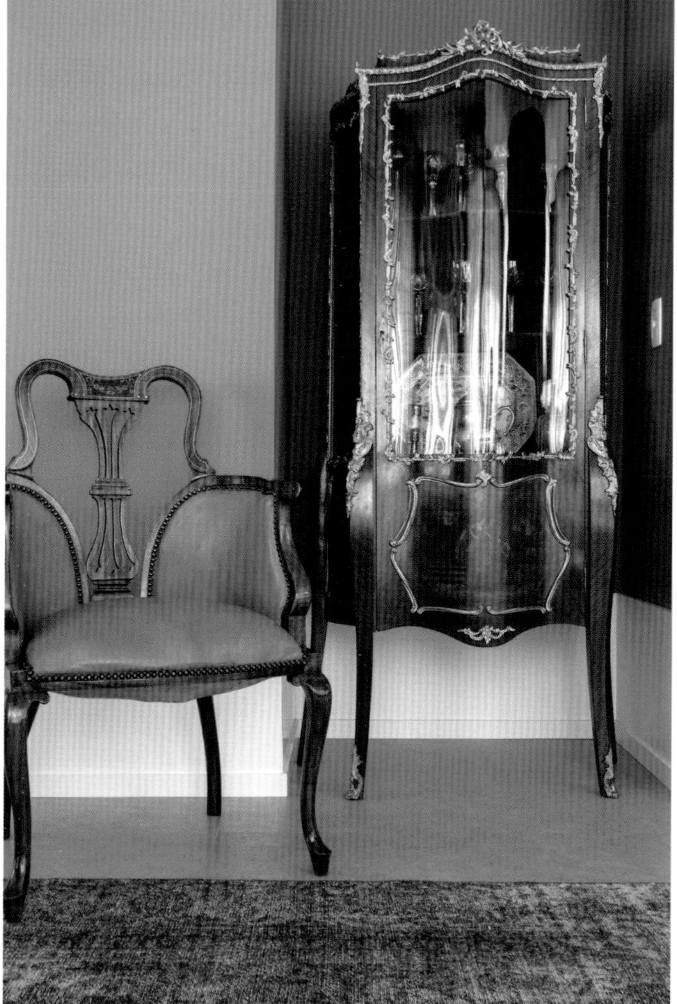

看来岂是寻常色，遍地花开映眼眸

As Eyesight Goes, There Is Always Flowers

项目名称：上海闵行星河湾 A1 样板房
设计师：邱德光

Project Name: Shanghai Star River A1 Show Flat
Designer: Qiu Deguang

在上海闵行星河湾A1样板房一案中，邱德光将奢华ArtDeco风格淋漓尽致地展现出来。室内从天花到地板均以几何的、纯粹装饰的线条来表现，如扇形辐射状的太阳光、齿轮或流线型线条、简洁的对称几何构图……地板大理石纹样对应天花弧形穹顶，从色调到层次渐渐递进开来，给人精致而又奢侈的感官享受。

公共空间设计讲究东方与西方意象交融，中式与西式家具混搭，构成一幅迷人风情图。花团锦簇的中式图案靠椅，雕画繁复的漆柜，加上各种时尚高档的西式沙发、灯具，各种物件混搭得宜，更彰显了居室的个性与独特。在色彩的处理上，邱德光选择了一贯低调的深棕色及更暗的色系搭配，又利用艺术品、灯具和陈设偶尔打破这种深色调，以实现空间视觉的平衡。

而对于单品的选择，邱德光大胆运用了多色系的家具单品，如同系出名门的贵族范思哲，让人眼前一亮而又并不腻味。ArtDeco的复古贝壳型钥匙盘搭配翠绿矮花瓶，正好衬托后面现代感十足的画作，色调丰富却又不凌乱。浅黄色复古座椅上的花团呼之欲出，让观者不禁感慨“好一树梨花压海棠的热闹春意”。凡此种种，单品的选择极尽优雅和奢华，也表达了邱德光对时尚、复古和精致的定位。空间内任何一件单品绝不只是随手的摆放，它们所处的位置以及体积都是为了使室内空间呈现无懈可击的软装状态。

在卧室的处理上，设计师不再沿用客厅的重合花纹地毯以及走廊的大理石几何花纹地板，而是选择了浅白蓝色与棕黄色调来彰显优雅恬淡的家居风格，给人更大的选择空间。卧室矮几上装饰的兰花草，象征着君子如兰的品质，也代表着“斯是陋室，惟吾德馨”的空间精神。从琳琅满目的会客室进入此间，有的不仅是身体的放松，更是精神的愉悦。

As a show flat, a project this space makes where to incisively and vividly make a luxurious interpretation of ArtDeco by Mr. Qiu Deguang. The ceiling and the flooring are done with geometric and decorative lines, like fan-shaped sunlight, gear, steam line and succinct symmetry of composition. The pattern of the marble flooring echoes with the arched dome of the ceiling, hues and gradation overspread progressively to allow for sensual pleasure of delicacy and luxury.

The public space lays emphasis on the combination of the east and the west. Furnishings eastern and western make an enchanting picture. Chinese-styled arm chair, carved and complex painted cabinet, sofa of different kinds, and lighting fixture compliment the unique personality. Nigger-brown as well as hues much darker are employed, but the dark stiffness is broken with art piece, lamp and furniture to achieve a balanced effect.

As for fitment pieces, more hues are involved, so the space looks refreshing but not monotonous or dull. Key disk is in form of retro shell type with short vase of jade green is complimentary to the modern painting behind. The flower seems to be coming out of the buff retro chair, commencing an image that the pear tree with flowers makes an overwhelming aesthetics that begonia can't. The utmost grace and luxury, no doubt, confides in the position of fashion, retro and delicacy. All items are not arranged at random, for their position and volume are just to achieve an impeccable decoration effect.

As for the bedroom, the patterned superposition of carpet and the geometric texture of the marble flooring are no longer continued, but hues of off white, blue and tan sets off the elegant grace of the interior, provide more choices for occupants. The orchid on the short table symbolizes the quality of gentleman, a representative of the spatial spirit that though it's a humble room, the owner's character is noble. The transition from the living room full of beautiful items into the comparatively simple bedroom is bound to relax and rejuvenate human soul and mind.

沁蓝轩朗，畅享生活的极致

Height and Broadness in Blue to Make an Ultimate Life

项目名称：融侨新城泷郡别墅样板房
设计公司：上合设计顾问有限公司
设计师：余周霖
参与设计：叶志应、王舟
摄影：三像摄建筑摄影机构 张静
地点：福建福清
面积：850 m^2
用材：树瘤木烤漆、皮革、镜面不锈钢、大理石、实木复合地板

Project Name: Show Villa of Rapids County
Design Company: Shanghe Design and Consultant Co., Ltd.
Designer: Yu Zhoulin
Participant: Ye Zhiying, Wangzhou
Photography: Threeimages Zhang Jing
Location: Fuqing, Fu Jiang
Area: 850 m^2
Materials: Stoving Varnished Burr, Leather, Mirror Stainless Steel, Laminated Solid Wood

独栋别墅应该是生活空间最为宽裕、舒适的户型结构，这样的空间就不再是仅仅满足于吃饭、睡觉等简单的家庭活动，而是以多样化的功能分区为居家生活提供了更大的可能性。现代便捷生活的痕迹在此套居室中展现得淋漓尽致，居家内部还配有电梯，方便主人活动，进入家中就等于进入了一个私人小世界。

一层作为公共空间，是主人招待客人的地方。客厅与餐厅比邻，客厅的挑高高达2层楼，装上华丽硕大的水晶吊灯，也丝毫不觉突兀。整面的落地窗，让空间显得格外透亮，而窗外的美景也自然地成了沙发的背景。客厅的基调十分简约，围绕茶几分布的沙发就是客厅的主体，水蓝的绒布沙发，在大理石背景墙的衬托下尽显优雅。石料是空间的主材，各种式样、颜色的大理石装点着空间，使空间愈发大气与沉稳。

居室内配有多间卧房，每间卧房都是一间设备齐全的小型套房，配备有休憩区、更衣间、大型卫生间，完善的设施让主人的生活更加舒适便利。不同的卧室则根据居住人群的不同再加以调整，打造成更为适合居住者气质的空间。酒吧、茶室、娱乐室一应俱全，招待一群好友也完全可以满足。整个居家空间秉承居者至上的设计理念，以利落、时尚的现代风格演绎出富有格调的居家气质，居者住的不再是单纯的普通居室，而是在享受一种全新的高贵生活。

The structure of a free-standing villa should guarantee that life can take place in a most comfort and spacious setting, which not only meets demands of food and sleep but that of more activities. Life easiness and convenience are available in this project, which is complemented by lift. Here really makes a private world.

The first floor serves as the public place, making a good place to receive guests. Adjacent to the dining room, the living room is as high as two floors, where the chandelier is gorgeous and magnificent but looks natural instead of towering. The French window covers the whole wall and compliments the spatial brightness. Views outside are taken into as an internal backdrop. The living room is simple, where velvet sofa of water blue dominates the space, bringing elegance and grace against the marble wall. Marble of different shapes and colors makes the staid and calm atmosphere become grand and generous.

Each bedroom completes a suite small but well facilitated with resting area, dressing room and bathroom. Life within is meant to be easy and convenient. Different bedrooms are kept individual for personal preference and taste. With bar, teahouse and entertainment room, small parties with friends can be held. The whole living space is neat, crisp and fashion. With living-oriented carried throughout, home life is not just so, but a completely-new life of dignity.

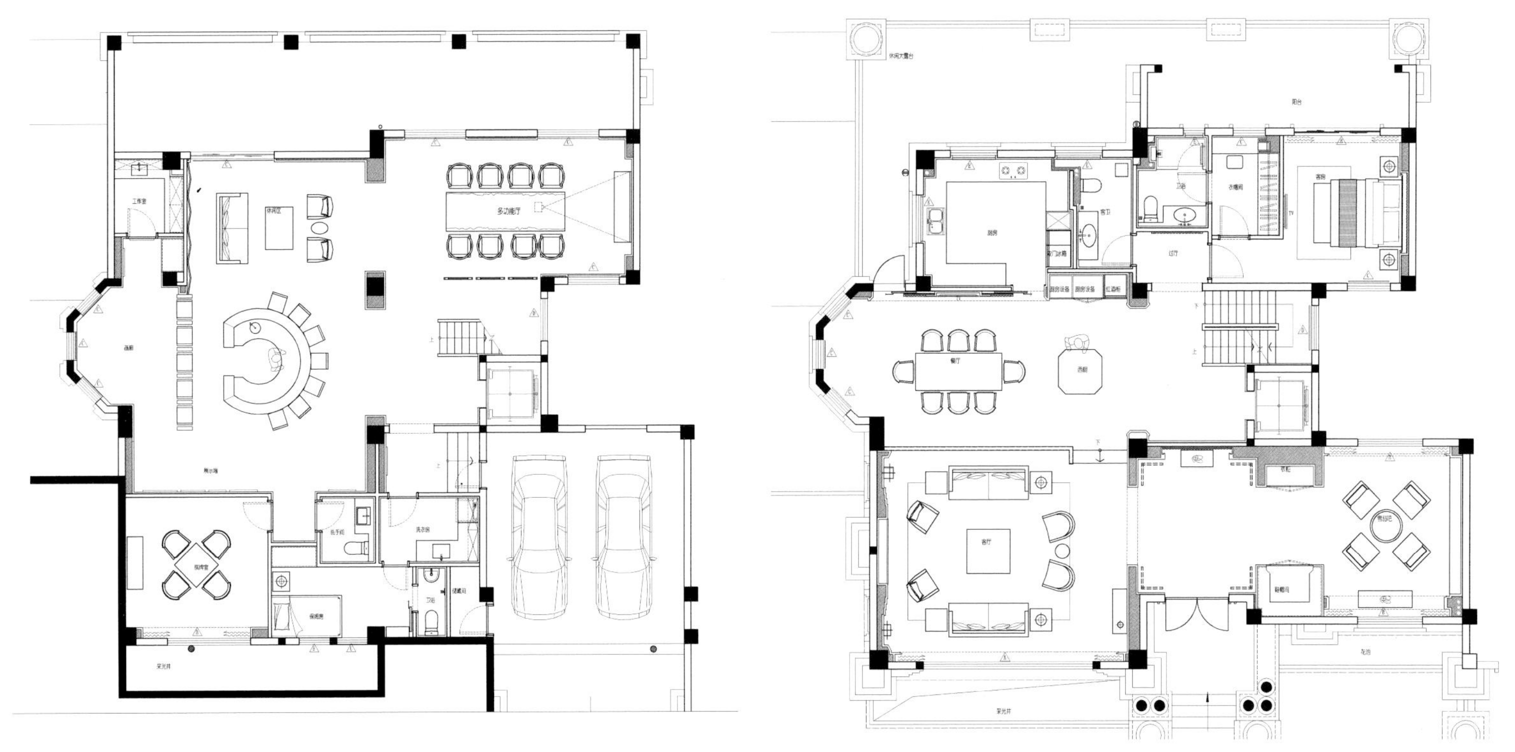

一层平面图 / First Floor Plan　　二层平面图 / Second Floor Plan

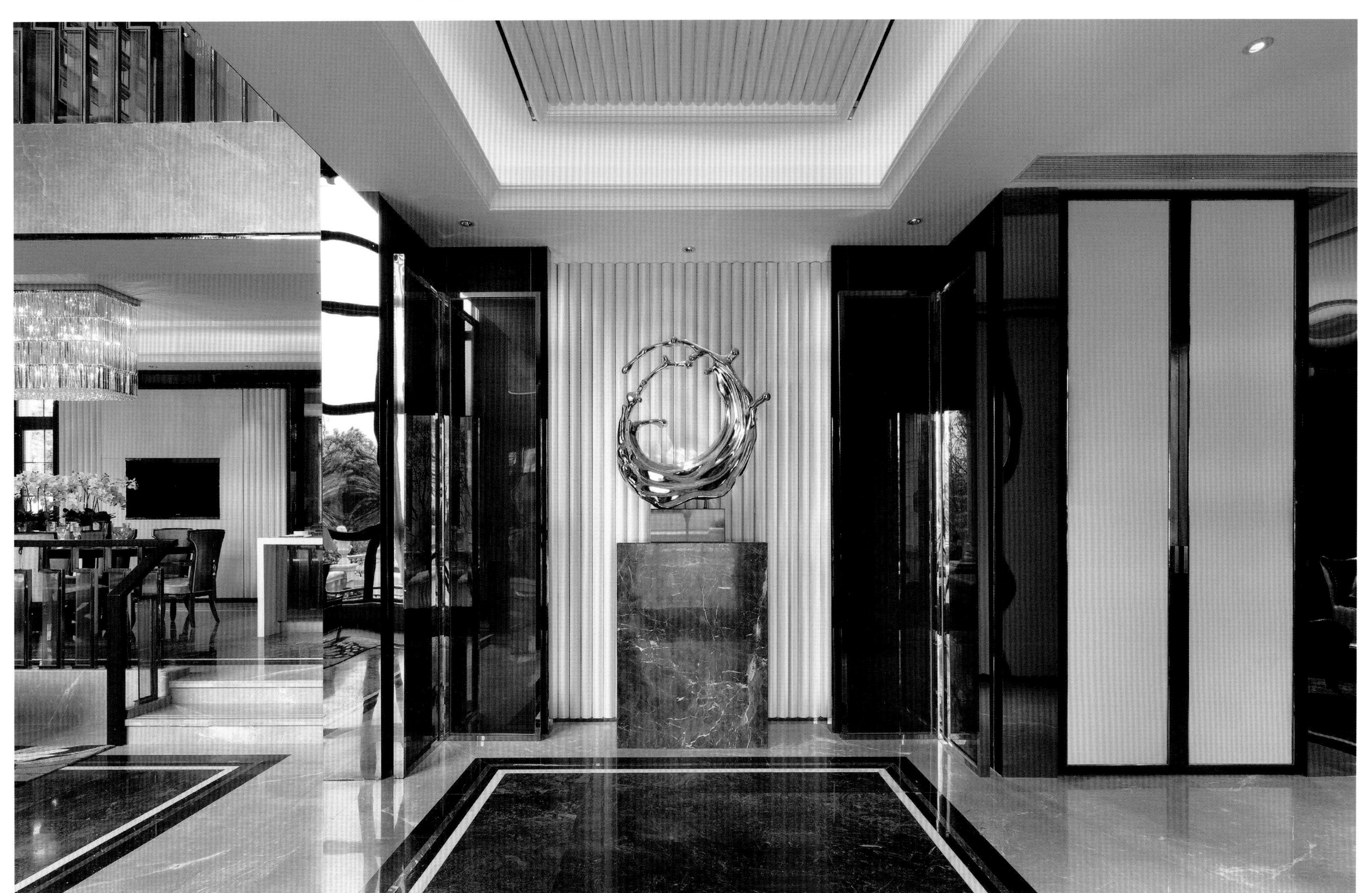

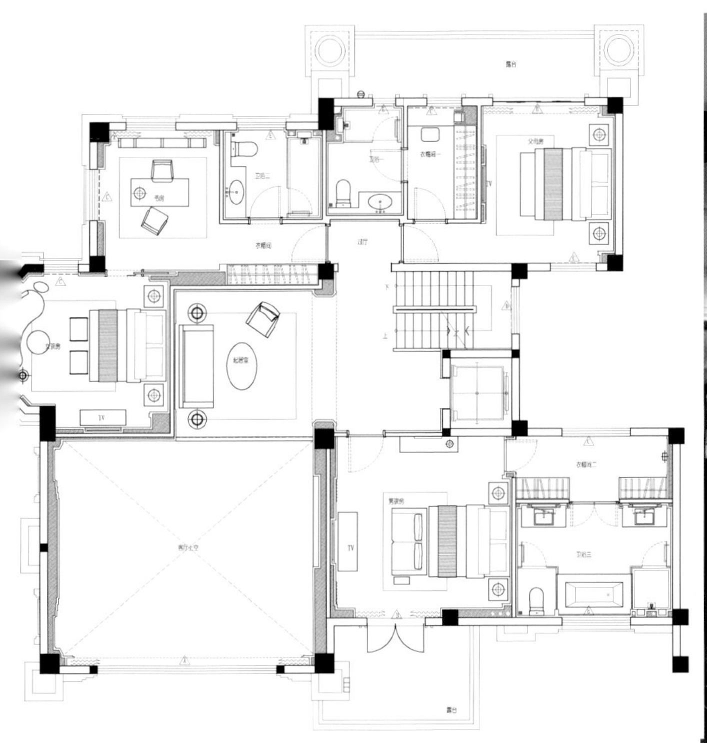

三层平面图 / Third Floor Plan

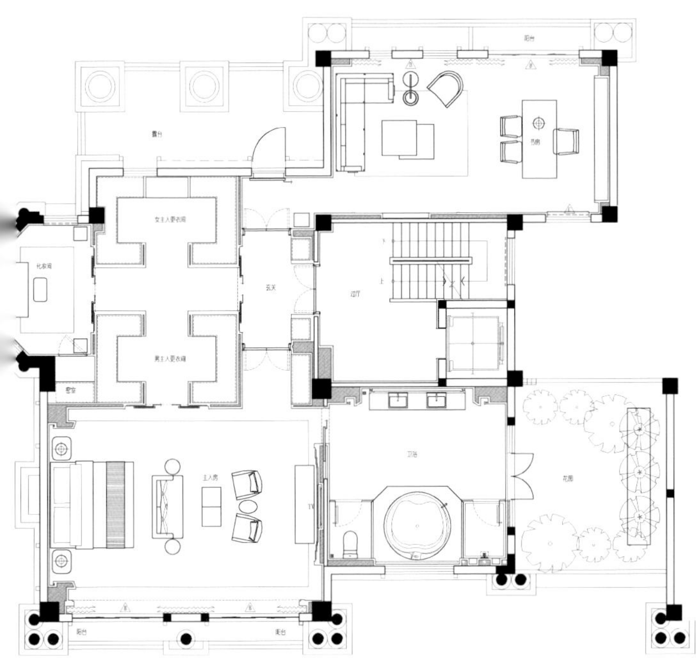

四层平面图 / Fourth Floor Plan

Eastern Imagination of French Neo-classic

法式古典的东方想象

项目名称：广州中信山语湖　　*Project Name: CITIC (Guang Zhou) Shan Yu Lake*

中信山语湖作为亚洲私家球场物业，历经多年悉心打造，成为一个集度假、休闲、商务、居住为一体的绿色国际居住新城。本案作为其样板房，设计师充分利用别墅大面积的优势，在布局上除家庭必备房间、客厅、厨房等设施外，还设有休闲厅、家庭影院等，使之成为一个集居住与娱乐为一体的大型住宅。

样板房设计每一处都充分彰显了本案独有的历史感与背景感，其风格来自拿破仑时期的新古典主义，并融入了许多现代材料，除去了古典主义繁琐的装饰细节，又融合进浓厚的中国元素，体现出气度非凡的大宅生活态度。

设计主要以“男人的权力”为表现内容，色调上以橙红与金色为主，铜为主要应用元素，沉淀出屋主的文化情怀。精简的古典欧式线条，使之成为适应新兴时代贵族的法式室内装饰设计，既延续了家族精神，又为之后的传承奠定了基础。所有家具装饰设计均为定制式，桌子、椅子、灯具等所用的铜都是100%纯铜，同时还结合了名贵胡桃木的包边，使得整个空间在体现沉稳大气的同时又别有一番优雅的气质。

As a ball park real estate, the project of Shan Yu Lake has made a green international dwelling community that integrates resort, leisure, business and residence. This space is a show villa, whose large area is made the best use of to accommodate necessary sections for living, including not only halls and kitchen, but also rooms for leisure and home cinema to offer a large home for living and entertainment.

Everywhere is employed with historical sense. The style is neo-classic from Napoleon period, which combines modern materials in leaving out classical trifles and blending a wealth of Chinese elements to embody an extraordinary attitude living in a mansion.

The design is to embody an idea of male power. Colors of orange and gold and element of copper are dominating the space to set off the personal culture taste. European lines brings forward a French interior design that's been widely popular among newly-rising nobles, carries forward the family spirit and lays a good foundation for future heritage. All furnishings and accessories are custom. Tables, chairs and lamps are of 100% copper, which with bordure of walnut boasts the spatial grandeur and elegance.

WESTERN

The Coexistence of Middle-Asian Charm and Classical Aesthetics

中亚风情与古典美学的浪漫共生

项目名称：杭州昆仑府 A1 样板房
设计公司：上海轩栾设计咨询有限公司
设计师：陈清芳
面积：660 m²

Project Name: A1 Show Flat of Kun Lun Mansion, Hangzhou
Design Company: Shanghai Xuanluan Design and Consulting Co., Ltd.
Designer: Chen Qingfang
Area: 660 m²

从伊斯兰文化中寻找恒河流域的皇室美学是本案设计的中心思想，本案为中亚地区住宅风格的局部缩影，融合了印度、斯里兰卡、巴基斯坦的伊斯兰古典美学。图案线板、木雕工艺的局部安排协调平衡，精致的大理石拼接传达了印度石材工艺的精美及殖民地时期的西方文化艺术，如泰姬玛哈陵的大洁白理石工艺细部设计被安排至各门套从风格来表现接近新古典的现代氛围，避免产生过于沉重的宗教感。

辅助建材则运用贝壳、有色玻璃、不锈钢、手工漆、真丝壁布，以及伊斯兰风格的各种面料来软化空间视觉，而配套家具则利用新古典的家具搭配印度风格的各种面料、皮革使家具仍具有主流气息及时尚生活的品位，软装饰品也选用摩洛哥及巴基斯坦风格的传统摆件、花艺来营造空间生活气氛，灯饰则以新古典的大气与印度小品灯饰并重。而空间的图腾主题串联则以印度女神像吉祥天的法印为全案精神，传达出本案空间设计的智慧、富裕、和谐、安全四大内涵，而这也是吉祥天女神在护佑人们所掌握的稳定能量，所以从入口的意象表现就能传达出印度三神一体的和谐象征。

入口处墙面挂件是典型英国皇室的礼宾烛镜挂饰，让来访的宾客未进之前玄关就能体会到极尽优雅贵气的华美开场。进入客厅迎面而来的除了吉祥天印的壁面主题装置镜饰外，更带点时尚气息。新古典元素的高光伊斯兰列柱及四方格木的主题墙。副饰面墙上的印度串珠宝石肃饰大画也突显了印度皇室细致的生活面向，天花顶景则由英式古典东方风格的水晶串珠大吊灯折射着让镶着云贝的顶饰，空间金碧辉煌。

而进入餐厅，映入眼帘的是以蓝碧孔雀标本饰物开创的伊斯兰盛宴，桌上的餐具器皿是由摩洛哥及印度传统食器组合并运用红色的色彩组合来营造具有中亚风格的用餐气氛。壁面的挂毯也传达了西方传统生活美学。二楼主要是以两间客房空间为主，空间以沉稳的咖啡亮光及胡桃木哑光来表现空间饰面的不同光影效果。三楼是主人的私密空间，此区分成两个块状安排，主卧室运用印度电影艺术宝莱坞风格的夸张色系来表达主人浓烈的生活情怀。主色是以新古典的黑高光饰面线条及有殖民地色彩的英式四柱床架来表现具有神秘古典及时尚风情的空间架构，而副色系就如星火燃烧般提亮整个空间，如玫红色、洋红色、深紫色、粉橘色。从床品配套、

窗帘纱幔落地大灯及花艺烛品都能体现出华丽精致的罗曼蒂克气氛，而主卫则是以斯里兰卡的白色基调为色系中心，其大理石细部手工雕塑的斜格纹中亚风的主题墙面传达了中亚艺术文明的精神美学。豪华大桑拿浴缸及池畔的香槟花艺的浪漫组合，使沐浴中也能享受不一般的贵族生活。而整个作品的精彩之处，莫过于连接各层空间的梯间设计，除了运用主题图腾元素吉祥天印的符号引导空间视觉之外，也大胆地运用了进口水晶手工漆，使楼梯扶手的黑色线条与孔雀蓝交错出具有印度风情的垂直片景，完美地传达了具有新古典的时尚及印度美学的伊斯兰风情。

DESTINY
DESTINY
Arletta
Arletta

The imperial aesthetics along the Ganges River is the design idea for this project, where dwelling style in Middle Asia is the partial epitome in fusing the classical Islamic aesthetics of India, Sri Lanka and Pakistani. Patterned line board and wooden carving employed in some sections serve as balance and are parqueted with marble to present the artistic delicacy of Indian stone process and the western art in its colonel period, e.g. details, including door pocket are treated the way the white marble is used in Tai Mahal just to get near to the modern atmosphere of the neo-classical in terms of style, avoiding successfully the heavy burden by rigidly adhering to religious art.

The auxiliary materials cover shell, stained glass, stainless steel, handmade paint, silk wallpaper and Islamic fabric to soften the spatial vision. Neo-classical furniture is accompanied with India fabric, and leather, furniture thereby kept a taste widely popular but of life fashion. Traditional Morocco and Pakistan accessories and floriculture are used to set off the life atmosphere. Lighting of grandeur as well as India lamps are simultaneously available. The dharmamudra of an Indian Goddess conveys wisdom, wealth, harmony and safety intended to be expressed in this project. That is what the Goddess, the Lucky Spirit holds. So the artistic concept symbolizes the Trimurti even from the entrance.

The protocol hanging drop of mirror and candle in the entrance, typically used in British royal family, allows for an utmost luxury and grace just at the beginning. The wall decoration in the living room is themed with the dharmamudra of the Indian Lucky Spirit that applied to decorate mirror offers some fashion. The glossy Islamic column, a neo-classical element and the check wall enhance the spatial magnanimous sense. The painting of Indian string of beads highlights the life in India imperial palace. Down the ceiling, hangs the bead-strung chandelier of neo-classical British style. With light from the chandelier, the shell ornament for the ceiling becomes more splendid and brilliant.

In the dining room, a sample of blue peacock starts an Islamic banquet. On the table is a collection of Morocco and traditional Indian feeder, the red onto which boasts a Middle-Asian dining air. The carpet on the wall detailed presents the western life tradition. The second floor houses two guest rooms, and rooms respectively for children and parents. The glossy brown and the matt walnut bring out the light and the shadow of the spatial facing. The third floor, a private castle for the owner, is divided into two blocks. The master bedroom involves a hue that's widely used in Bollywood films for presenting the strong feelings of the owner. Besides the dominated hue which is accomplished by neo-classical glossy facing and British colonel canopy bed, berry-red, magenta, dark purple and pink-orange are also used. Items, like bedding, curtain, mantel, floriculture and candle, give expression to a gorgeous and delicate romantics. The master bathroom is coated in white of Sri Lanka and fixed with a marble check wall to complete the aesthetics of the Middle-Asian civilization. The traffic line is centered on the sauna bathtub and the bubbly floriculture around it, bound to give enjoyment only exclusive to the noble. The most remarkable is the staircase to join the whole space, where in addition to the dharmamudra of the Indian Lucky Spirit, the crystal handcraft paint with the peacock blue shifts lines of the stair into vertical scenes of Indian style. This perfectly delivers a combination of the neo-classical fashion and Islamic aesthetics in India.

A Space of Tai Ji: To Simplify Complication

以简驭繁，太极空间

项目名称：佛山水悦龙湾 61 号别墅
设计公司：深圳雅典居设计
设计师：陈昆明
面积：500 ㎡
用材：石材、瓷砖、不锈钢、皮革、布艺

Project Name: Villa 61 Dragon Bay, Foshan
Design Company: Shenzhen Hover-house Project Design Co., Ltd.
Designer: Chen Kunming
Area: 500 m²
Materials: Marble, Tile, Stainless Steel, Leather, Fabric

一座宅邸的居住品质如何，很多时候取决于业主和设计师的审美品位。此次佛山水悦龙湾 61 号别墅通过时尚高雅的现代简约风格，突显了其作为一座豪华宅邸应有的品质。它摆脱了一般豪宅的传统造法，以中性的灰调子代替金碧辉煌的色彩，调和出一种沉稳的韵味。它不像女人的服饰那般珠光宝气，闪烁夺目，相反，它就像是一种气质的积淀，和谐、恬静、脱俗于众。而这，就是设计师赋予它的一种生命的从容与悸动，给予心灵些许冲击。

别墅囊括了客厅、餐厅、卧室、书房、娱乐室、藏酒区等区域，功能齐全。客厅在石材、瓷砖、不锈钢、皮革、布艺等的建置下，透过光晕的反射，让空间微微呈现出多重质感与景深，幻化出空间美感。餐厅与厨房摒弃不必要的浮华，把现代简约风格发挥到极致。卧室细腻简约，运用灰色与白色糅合出空间韵味，辅以间接照明的柔和灯光，让睡眠空间散发优雅的气氛，更强调了居家舒适感及内心的满足。藏酒区井然有序地排列着装满藏酒的酒架，采用不锈钢制作而成，时尚潮流，远远看去，犹如书店中一排排塞满了书籍的书架，让人一时分不清到底是酒香还是书香。

为了强调空间的艺术氛围，设计在软装方面适当铺排不同的艺术摆件与画作，以此将一种独特的艺术气息散发在空间的各个角落，陶冶性情。在这里，色彩沉稳的各种布艺搭配园艺花木的配置，让居者和访客更多地感受到一份儒雅舒适的格调与随心所欲的自在。

The quality of a house is frequently determined by aesthetic taste of the owner and the design. The project of Villa 61 highlight qualities a luxurious mansion should have when positioned as minimalist but fashion and graceful, quite differing from the usual luxury by being endowed with a grey netural tone more than resplendent and magnificent to set off calmness and staidness. Not only the jewel on fashion, it's of harmony and tranquility, and refined. And this is the leisure and power to allow hearts impact beneath the skillful hands.

A fully-functioned this space is to have accommodated rooms of living and dining, bedroom, study, entertainment, and cellar. In a setting of marble, tile, stainless steel, leather and fabric, the living room brings out layers and scenes coming one after another. The dining room and the kitchen have abandoned the fabulous necessity by giving utmost play to the minimalist. The bedroom with grey and white is simple and delicate, where indirect lighting compliments the good sleeping ambience. This on the contrary stresses the comfort and inner satisfaction. Around the cellar are wine racks of stainless steel, which, if looked far away, are like book shelves filled with books.

The employment of art pieces and paintings is overspread everywhere for creating artistic ambience. The configuration of fabric of variety and flowers and trees, make the occupants and visitors feel nothing but comfort, free and gentle.

PRESLEY
HEPBURN

阿拉伯豪宅
Arab Mansion

Memory of Alhambra

阿尔罕布拉宫的回忆

项目名称：苏州中海·鸿堡
建筑及景观设计：UA 国际
室内设计：水平线设计
面积：50 600 m²

Project Name: CSCL (Su Zhou) Fortune Castle
Architecture and Landscape Design: Urban Architecture
Interior Design: HSD Horizontal Interior Design
Area: 50,600 m²

鸿堡以阿尔罕布拉宫为主题来设计，位于苏州独墅湖南侧独墅岛上。通过对苏州城市肌理和脉络的梳理与解读，设计以巷、院来组织各个单元空间，由巷进院，由院入户，构建具有苏州特色的空间肌理。采用整体地下车库进行人车分流，满足现代人生活需求的同时，营造一种过巷入门庭，进院落的空间意境。

鸿堡没有复制一个阿尔罕布拉宫，而是用现代手法重构了一个追忆阿尔罕布拉宫的场所。鸿堡将不同文化背景的建筑很好地移植到当代生活，营造出适合现代人居的带有异域文化特征的全新空间，也是一种很好的设计尝试。

鸿堡试图从院落、水、光影、装饰、材料等几个方面重构阿尔罕布拉宫的回忆。

院落是中国人传统生活方式的载体。鸿堡即是参照阿尔罕布拉宫院落特点，通过廊子、围墙、照壁，以及一些非建筑元素如植物、雕塑等界定领域，形成玄关、主院、后院、下沉庭院等大小不一或私密或开放的多层次空间结构。

阿尔罕布拉宫对光有巧妙的运用。为达到相同的效果，鸿堡刻意设置了一个与玄关相连的前厅，途经微暗的前厅突然迎来庭院耀眼的阳光，光亮和黑暗的反差带来错愕之后的惊喜。除了摩尔式装饰，鸿堡采用了大量当代材料来诠释传统意境。在这里嵌有防腐木的工字钢替换了阿尔罕布拉宫白色大理石的柱子，原本复杂的横饰带被简约的工字钢所代替，铺砌有绚丽的釉面砖的壁脚板及墙身也替换成了现代材料劈开砖和真石漆，虽少了奢靡之气但不失现代的高贵典雅。

水是阿拉伯文化中极为重要的元素。水的运用使鸿堡有了一种灵动之感。

157

Fortune Castle is a space with a motif of Alhambra. The project, located on Dushu Island, south of Dushu Lake makes spatial texture of Suzhou by involving the local urban vein and planning sections inform of lanes and courtyards. The parking lot underground realizes the complete traffic independence of people and vehicle, not only meeting the needs of modern life but creating a spatial artistic concept of walking through lanes and into atrium.

Here makes a space in memory of Alhambra with reconstructing approaches instead of by copying to make another Alhambra, where to make a good trial to explore design by transplanting buildings of different cultural backgrounds into contemporary life, completing a new space for modern living with exotic cultural identity.

All, including employment of courtyard, water, light, decoration, and materials, renew the old memory of Alhambra.

Courtyard is a carrier to accommodate traditional Chinese lifestyle. Fortune Castle is characterized by reference to the Alhambra courtyard, where through the definition of porch, fence, screen wall and some non-architectural elements like plants and sculpture, multi-level structures come open or closed, big or small, of the entrance, the main yard, backyard, and the sinking garden.

As for light, it's used cleverly in Alhambra. To achieve the same effect, a vestibule connected to the entrance is deliberately set up. Through somber vestibule, sunshine is cast bright suddenly into the courtyard. The bright and dark contrast offers a stunning surprise. In addition to Moorish decor, here is decorated with a lot of contemporary materials. Joist steel embedded with wood preservative-treated timber replaces white marble pillars in Alhambra, when taking place of all complex. The skirting and the walls are of split brick and coated in lacquer instead of being coated in colorful glazed tiles, which, although less extravagant, look with more modern elegance.

Water is an extremely important element in Arab culture, the use of which here brings forward ethereal feelings.

Arabic Fantasy with Modern Approaches

现代手法，演绎阿拉伯的绮丽风华

项目名称：大别墅
摄影师：大卫·布兰克

Project Name: Villa Majestic
Photographer: David Blank

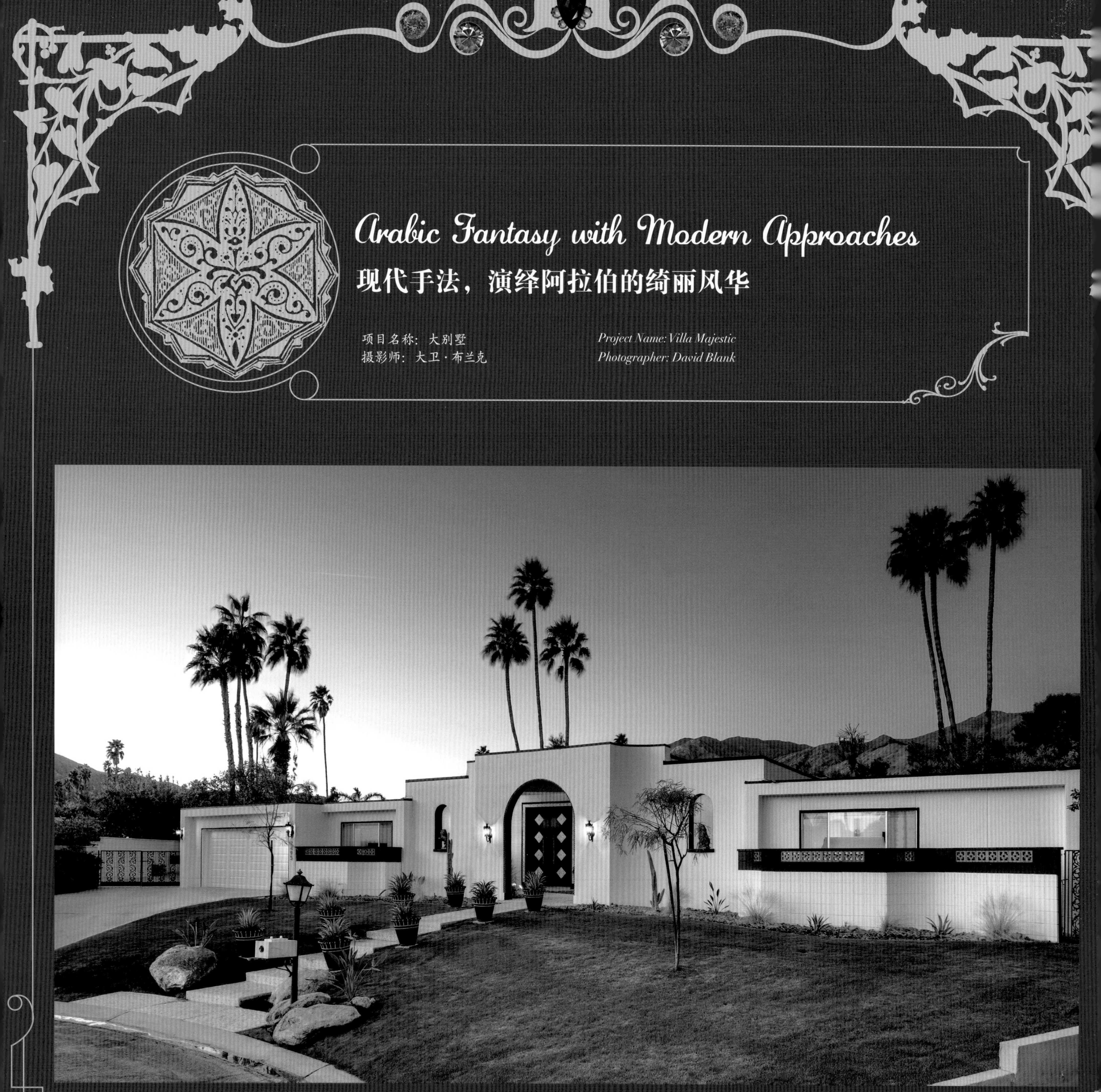

本案原本早已是破败不堪，经重新设计整修后，空间内外焕然一新。设计的灵感丰富多彩，些许源于电影《*Topper*》《*Thin Man*》，些许源于20世纪30年代的喜剧，还有一些理念则源于著名的装修大师比尔·海恩斯。

室外以黑、白、金为主调，照亮着原本不同寻常的立柱、门窗、拱顶及平面的屋顶。

室内延续着黑色、金色的主题，为墙面创造出一种维度及戏曲化的张力。拆除了玄关、大厅、客厅之间的墙体，诺大的空间成了居家生活的娱乐天地。宽大、低矮的绒布卧式沙发覆盖着银白色的金属皮质。

五个卧室的灵感源于名牌寝具。既各成世界，又相互融合，明亮、传统但却与众不同。不同的卧室，不同的床头设计。专门恢复的水泥地板专为空间增加了一份久远的沧桑感。

厨房里新添了白色乙烯基的凉亭元素，饭店的风格尽在其中。凉亭后面的墙体也因此得到了升华。整个空间曾经的画廊质感也因此得到了进一步提升。

室外，泳池周边的水泥早已是破败不堪。新换上的混凝土板，在传承的同时，又给人几分闲适感。除了黑白条纹的更衣室、SPA、烧火坑无不与房子的屋顶轮廓线相辉映。

This is a project refurbished from a very distressed property, which makes a success by removing all coverings inside and out. Its inspiration for this house comes from the movies, such as TOPPER, THE THIN MAN, and many other high style comedies from the 1930's as well as from famous decorator Billy Haines.

The exterior is endowed with a black & White & gold theme for, highlighting the unusual columns, ocular openings, arches and flat roof design.

The designer brought the outside in with using black and gold paint to create dimension and drama to the walls. Having taken down the wall between the entry, hall and living room, dweller had a grand space to play with. And the designer designed large, low, and deep, tufted sofas in Silver Metallic leather.

For the five bedrooms in this house, designer took inspiration from Trina Turk's home bedding line, and each room has a unique, yet highly unified feeling to it. Very bright, mod, and different. He designed different head boards for each room, and he restored the concrete floor through most of the house.

In the kitchen, he enhanced the original galley space by adding a restaurant style white vinyl booth, and accenting the wall behind it orange.

Outside, the concrete surrounding the pool was all broken and jagged. Designer removed the existing concrete, he found the great black and white striped cabanas and poured concrete pads to set the loungers on. He also added a spa and fire-pit that mimic the shape of the roofline around the house.

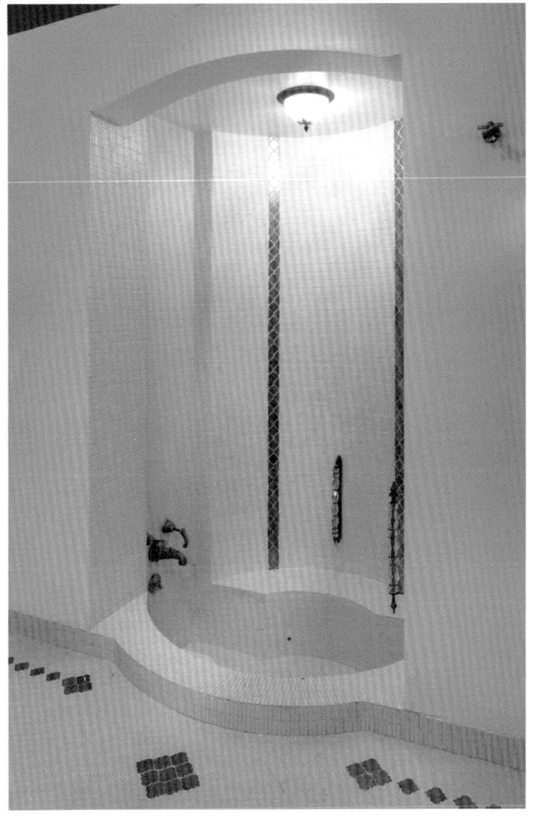

生活的一千零一夜传奇

The Thousand and One Nights in Life

项目名称：杭州金地自在城境墅别墅
设计公司：IADC 涞澳设计有限公司
设计师：潘及
摄影师：金沾
面积：450 m^2
用材：STUCCO、大理石、浮雕、仿古砖、木饰面、手绘壁纸、银、贝壳、水晶、丝绒

Project Name: Gemdale (Hangzhou) Free City
Design Company: IADC Design
Designer: Eva Pan
Photographer: Jin Zhan
Area: 450 m^2
Materials: Stucco, Marble, Relief, Antique Tile, Veneer, Hand-Painted Wallpaper, Silver, Shell, Crystal, Velvet

迪拜几乎成了奢华的代名词——世界上第一家七星级酒店帆船酒店、全球最大的购物中心、世界最大的室内滑雪场、源源不断的石油、巨大的财富……然而作为阿拉伯联合酋长国人口最多的酋长国，迪拜依然处处散发着浓郁的阿拉伯气息，神秘、浪漫、梦幻。设计师 Eva 首次挑战阿拉伯迪拜风格，试图以迪拜当地特有的文化元素以及阿拉伯皇宫的尊贵气质，打造一处充满童话色彩的奢华居所。

步入空间，强烈的阿拉伯风就迎面而来。阿拉伯建筑中最典型的拱门元素在室内兼具功能性和装饰性，成为空间中最为瞩目的设计亮点。设计师 Eva 介绍说，区别于室外的拱门，她将象征富贵堂皇、吉祥如意的孔雀引入设计中，对孔雀羽毛进行了简化，赋予拱门羽毛组合似的造型，使得空间表现更为柔美和尊贵。这种独特的设计元素随后在客厅、书房、浴室、卧房等处都频繁出现，不管运用于空间隔断还是壁龛装饰中，都是视觉的亮点所在。

阿拉伯装饰风格以各种植物和抽象曲线互相盘绕构成基本图案。在本案中，图案的运用也是营造风格的浓重一笔。拱门上镂空的雕刻装饰、天花吊顶的组合排列、公共区域地面的优雅瓷砖拼贴，几何纹饰、植物纹饰、阿拉伯书法纹饰或独立或组合出现，带来充满异域风情的空间印象。即便是最难表现和演绎风格的现代厨房和浴室，也因为这些特别的纹饰而变得与众不同，让人过目不忘。

伊斯兰建筑的颜色基本以米白米黄为主，Eva 在设计中也本着尊重阿拉伯精神的原则，以米白、米黄为主要背景色调。阿拉伯世界的浪漫与自然，对阳光、沙漠、骆驼等美好事物的感知，在色彩的运用中被表现得淋漓尽致，给人一种纯净至美的精神境界。而延续孔雀的装饰细节，室内大量运用孔雀蓝，纯净之上更增添了非凡、华贵、高雅的气质。而 Eva 特别为本案挑选的灯具、银器、织品、陶瓷、家具等，均来自中东地区，采用气质张扬的造型和装饰，加上金色、咖啡色、砖红色、紫色等的渲染，更增添了阿拉伯皇室的尊贵和奢华。

阿拉伯，这个古老文明的发源地之一，让人想起无边的沙漠、奇幻的海市蜃楼、长长的驼队、清脆悦耳的驼铃、身着白色长袍的领驼人、庄严美丽的清真寺、悠扬的诵经声、神秘的阿拉伯皇宫、一千零一夜的奇幻与华丽……设计师 Eva 的精湛演绎将如此逼真的阿拉伯世界带进了生活，同时加入迪拜现代生活的舒适与惬意，让人感受到极致的浪漫与奢华，回味无穷。

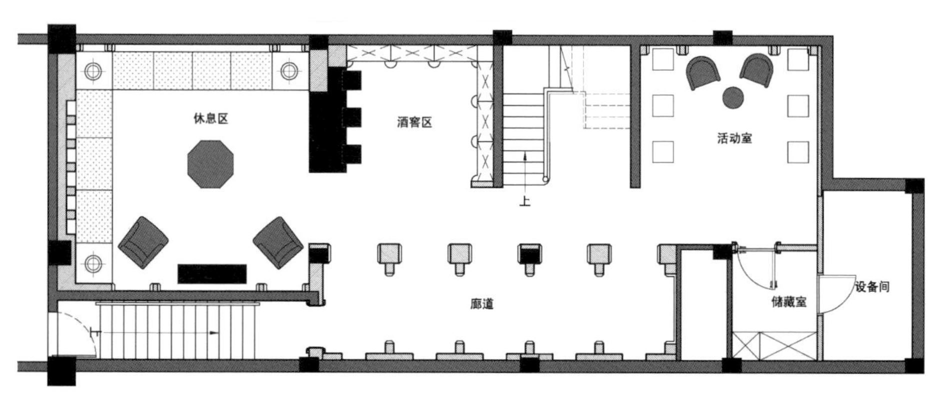

地下层平面图 / Basement Plan

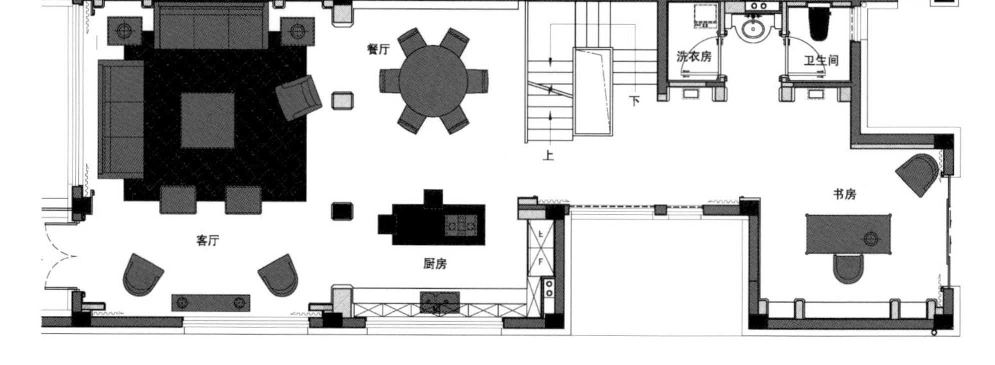

一层平面图 / First Floor Plan

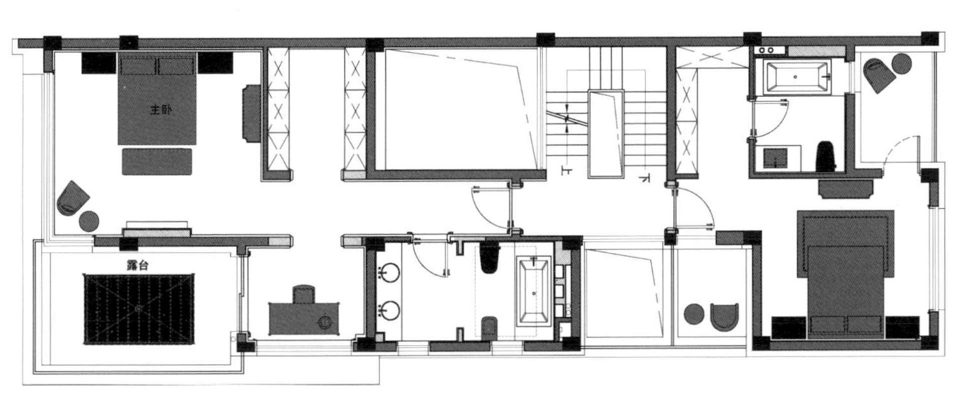

二层平面图 / Second Floor Plan

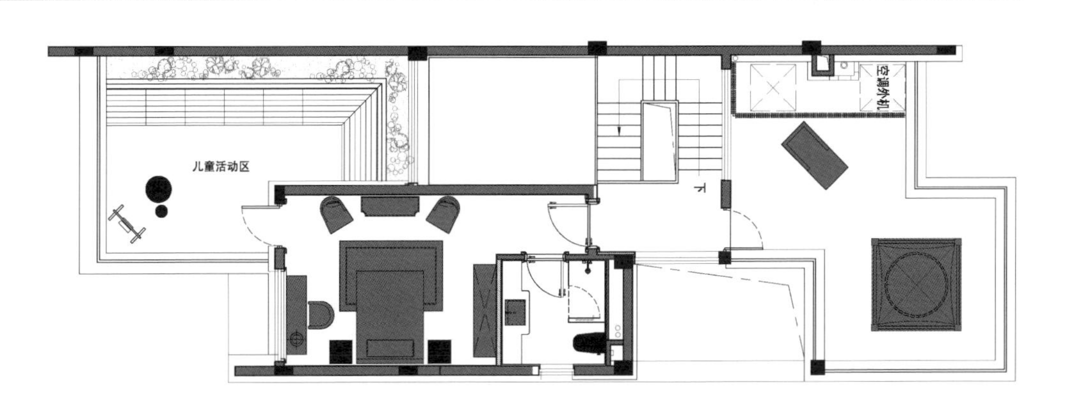

三层平面图 / Third Floor Plan

With the wealth pouring in rows with its abundance of oil, Dubai, almost a pronoun of luxury, has Burj Al Arab Hotel, the first seven star hotel in the world, and the largest shopping mall and the largest indoor ski slope in the world. As one with the largest population among the seven emirates of United Arab Emirates, Dubai is a place that is rich in Arabia ambience, mystery, romantics and dream. And this space makes a project that challenges Arabic style by Eva, who employs cultural elements from Dubai and dignity of Arab imperial palace to create a dwelling place of luxury, which is filled with the flavor of fairy tale.

The foyer features a strong sense of Arab. Arch, the most typical element in Arab construction that can be functional and decorative, completes the most eye-catching design highlight. According to Eva, the difference the interior arch makes, quite different from that outside, lies in its introduction of peacock. Peacock is a symbol of wealth and rank, and good luck and happiness. Peacock feathers are simplified here, so the arch is feather-like, the space thus being of more morbidezza and nobility. Such a design is available in many places, like the living room, the study, the bathroom and the bedroom. Whether used for partition or niche decoration, all are visual bright spot.

Arab decoration style consists of basic pattern of interweaving plants and curves in their abstract senses. And pattern employment plays a very important role in this project. The hollowed-out work on arch, the suspended ceiling, and the tile collage flooring in the public space are all in form of emblazonry, geometric, botanic and Arabic calligraphy. Whether kept independent or combined together, all are exotic. Even the modern kitchen and the bathroom as the place hardest to interpret the style, become unique and impressive instantly because of such an ornament.

Islamic buildings take off-white and cream-color as the principal hue. A principle that has been stuck to by Eva in this project. The romantic of Arab and the good perception of sunlight, desert and camel are expressed incisively and vividly, allowing for a spiritual purity and an utmost beauty. The peacock blue used in large amounts continues the peacock element in complimenting the temperament extraordinary, noble, and refined. The careful selection by Eva, like lamps, silverware, fabric, ceramics and furniture, are from the Middle East, whose bold modeling and decoration rendered with hues of gold, coffee, brick-red and purple add more of the nobility and luxury of the Arab imperial palace.

Arab, one cradle of the ancient civilization, is reminiscent of desert, flyaway, camel team, camel bell, and camel-leading Arabian in long robe. In this project you feel nothing but leisure and comfort of modern Dubai, and the paramount romantics and luxury that comes out of Masjid, sounding by reading classics and imperial palace. All are worth endless aftertastes.

To Come Here, to Enter Mystery of Arab

神秘的阿拉伯，走入梦幻秘境

项目名称：东滩花园
设计公司：壹陈设计

Project Name: Garden of East Beach
Design Company: Yi Chen Design

阿拉伯风格在室内设计中的运用，往往结合伊斯兰文化，展现阿拉伯世界的艺术与文化、浪漫与自然，让人对阳光、沙漠、骆驼等美好事物有所感知，带给人一种心灵粗犷豪放和精神纯净至美的感受。对于本案，设计师希望通过阿拉伯风格来展现高端物质层面以外的，能投射出更多精神内涵与心灵寄托的、与人们内心世界相契合的家。

设计汲取了阿拉伯建筑艺术精神，冠以多种形状的阿拉伯穹顶，并在穹顶上的天花板上织入不同的图案，它们酷似一张张透明的网，让光梦幻般的光弥散在室内。软装设计中，如水的蓝，如同精灵般映射于室内的每个角落，营造出舒心的居室氛围。

客厅地面铺陈米黄色的地砖，配以黛色、孔雀蓝的沙发，墙面的阿拉伯装饰画则对视觉感受起到了调和作用。相邻的餐厅内米白色橱柜、孔雀蓝的餐桌椅让整个空间不显沉闷，为生活带来清新之风。地下室黛色天花和浴室的水滴蓝天花，则让人如同沉浸在天空与海洋的世界之中。

设计赋予本案空间新的生命与灵魂，其区域划分明晰又不失整体性。通过不同的穹顶及天花予以功能空间的区分，如粗放却韵味十足的客厅、几何感鲜明的卧室。同样，空间布局、门、墙壁上无一不引用阿拉伯的艺术元素，具有“世界纹样之冠”之称的伊斯兰的纹样随处可见，让人无处不感受到深厚、神秘的伊斯兰文化艺术气息。

If used into the interior design, the Arabic style is usually combined with Islamic culture to present art, culture, romantics and nature exclusive to the Arab for creating perception of sunlight, camel and desert in human minds. The result is to offer mind openness and soul purity. And this project makes good reference to such a style for conveying a home where people can find a refuge for their heart that's in line with their inner needs, as well as high-end material needs.

Arabic domes of various kinds involved here are added with different patterns, which looks like a transparent net to ooze dreamy light into the interior. This carries forward the essence of Arabic tradition. Into the upholstering is interspersed with blue. Such a water-like color feels as if elves were everywhere to make a comfortable air.

The beige flooring in the living room is matched with black and peacock blue. The Arabic painting on the wall allows for a visual effect. In the adjacent dining room, off-white cabinet, pea-cock dinging table and chair makes here free of dullness, where life is refreshing. The basement is fixed with black ceiling; the ceiling is of water-drop blue. Here makes a world of sky and sea.

The holistic style is endowed with a new design concept. The definition is clear but not short of integrity. Various forms of ceiling and dome are now used for spatial partition. The living room appears rough but very romantic, and the bedroom is of a strong geometric sense. Artistic elements are overspread everywhere, like door and gate, so does Islamic emblazonry that's honored as "Pattern Crown of the World". All provide you nothing but an artistic sense of the Islamic mystery.

生态大宅
Eco-Mansion

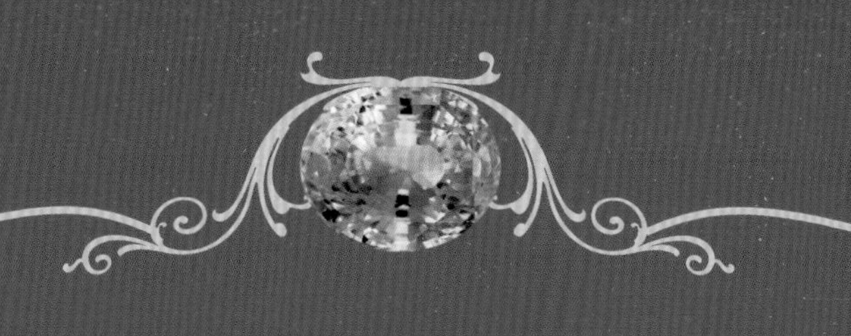

Return to Nature
回归自然

项目名称：圣桑别墅
设计公司：亚历杭德罗·桑切斯·加西亚建筑师事务所
摄影师：海梅·纳瓦罗
地点：墨西哥
面积：860 m^2

Project Name: San-Sen House
Design Company: Alejandro Sanchez Garcia Architects
Photographer: Jaime Navarro
Location: Mexico
Area: 860 m^2

圣桑别墅位于墨西哥一林地中央，为金属、玻璃、木混合结构。单层量体，内含三个房间，客厅、厨房、露台、阳台皆为木质梁柱天花。模块化设计，使各空间与自然环境良好衔接。

设计极尽空间本质，栖息之可能。结果简单、直率、永恒……

Situated in the middle of the Valle de Bravo forest, San-Sen House is a stilt house with a metallic structure incased in wood and glass. The single storey house contains three rooms, living area, kitchen, several balconies and terraces under wood beamed ceilings. Its modular design gives every space of the house a visual connection with the natural surroundings.

As the designers first approach to the design, they exhaust every possibility in order to resolve each aspect of the design. They are not trying to be original, on the contrary, most of the times the results are very simple, straightforward, and timeless...

Innovative Interpretation by Wood and Stone

木与石的创新演绎

项目名称：多伦多大公馆
建筑设计：Belzberg 建筑师事务所
建筑师：科里·泰勒、克里斯·阿恩岑等
室内设计：MLK 工作室
摄影师：本·拉恩
面积：837 m²

Project Name: Toronto Residence
Constrution Design: Belzberg Architects
Architect: Cory Taylor, Chris Arntzen, Brock DeSmit, David Cheung, Barry Gartin, Aaron Leppanen
Interior Design: MLK Studio.
Photographer: A-Frame / Ben Rahn
Area: 837 m²

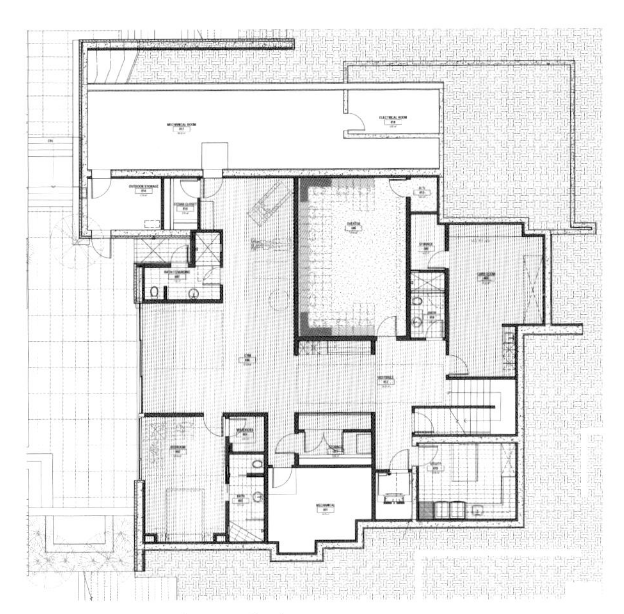
一层平面图 / First Floor Plan

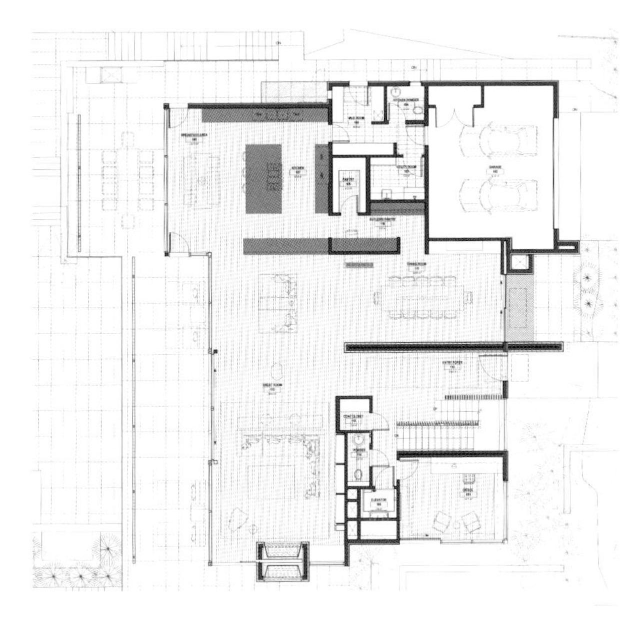
二层平面图 / Second Floor Plan

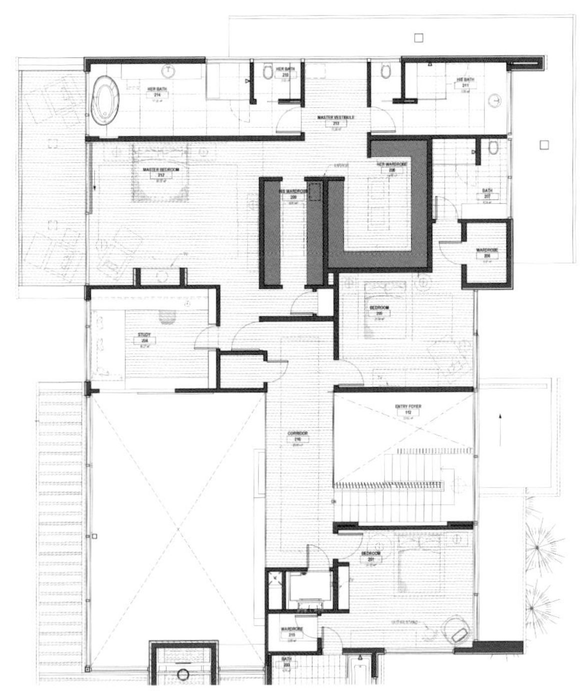
三层平面图 / Third Floor Plan

本案位于多伦多北部的顶级住宅区。设计师希望本案不仅能充分利用基地周围丰富的自然景观，更能成为国际家庭的枢纽。大面积的釉面区域、清爽的线条、简约的体量比例既彰显着业主的兴趣所在，又模糊了内外之间的界限。精心设计的内部布局，轻松温馨的家具饰物，低调之中却显个性。借助设计之匠心，建筑与景观设计赋予了空间丰富的自然光线与良好的通风，但隐私依旧独好。简单、持久的色调搭配着石膏、锌覆层，以及各种天然的木质与石材。

Designed for a large double lot in a premier neighborhood in north Toronto, the project's ambition is split between providing a space that can take advantage of the site's abundant natural features and also serve as a hub for a growing international family. Large portions of glazing along with clean lines and simple volume proportions underscore the client's interest in creating a space with an effortless flow between interior and exterior; in a climate that is not always conducive to this type of habitation. The interior layout of the house has been carefully crafted to provide an informal and inviting space with an understated sophistication. The architecture and landscape design at the front of the building have been choreographed to allow for an abundance of natural light and a feeling of airiness without sacrificing privacy. A simple, yet enduring material palette blends Plaster, Zinc cladding, along with various natural woods and stones.

平面图 / Site Plan

立面图 / Elevation Drawing

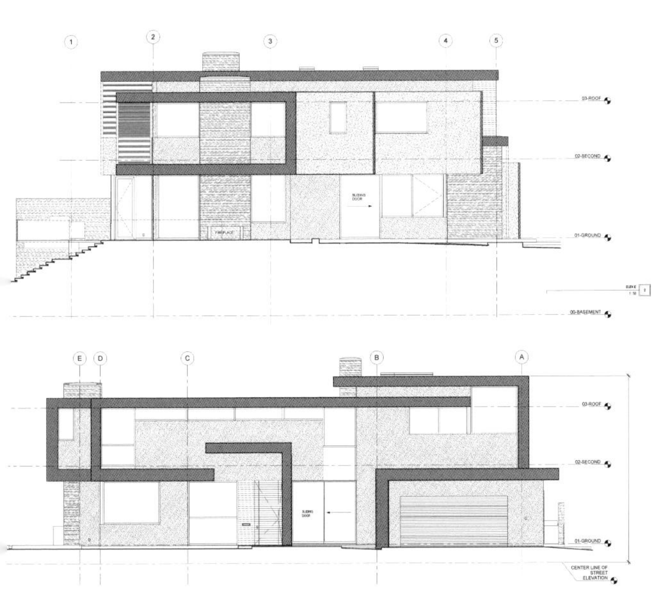

立面图 / Elevation Drawing

Enjoy Pastoral with a Humble Heart

以谦和的姿态回归田园

项目名称：怀劳谷别墅
设计公司：彭逊逊建筑师事务所
摄影师：保罗·麦克雷迪

Project Name: Wairau Valley House
Design Company: Parsonson Architects
Photographer: Paul McCredie

“怀劳谷”别墅位于德国，距布伦海姆仅10分钟的路程，而且还靠近怀劳河。放眼望去，满眼皆是农田、果园、葡萄园，好一派田园景象。

本案不仅在空间上，而且就用材而言，坚持彰显地方特色。不同地方的亭台楼阁，有开放，也有隐私，给人一种露营的感觉。

弯弯曲曲的山形墙，模糊着内外之间的界限。粗线条的水泥墙结构虽然界定着空间，但却使建筑如同自然天生一般。三角形的屋顶给人一种庇护的感觉，也为家居生活带来了舒适。山水风景在此得到了精心的呵护。成排的果树与葡萄藤架以其规整的图案标志着所在。行行绿廊、立柱与别墅的线条相互呼应，大有随时纳空间于其地的态势。

The property is located in the Wairau Valley, 10 minutes from Blenheim and close to the Wairau River, a patterned rural landscape of farms, orchards and vineyards.

The House has a strong connection to the land spatially and materially. There is a sense of encampment, where different interlinked pavilions offer different areas of occupation and privacy or openness.

The spaces are housed beneath low sloping gable forms where boundaries are blurred between the land and building. Heavy lined concrete walls and structures feel part of the land and are used to define indoor and outdoor spaces. The gable roofs float over these to provide shelter and the comforts of home. The landscape is cultivated; lines of orchard trees and vines mark the land in regular patterns and the house responds with the lines of pergolas and posts ready for planting to integrate the house with the land further still.

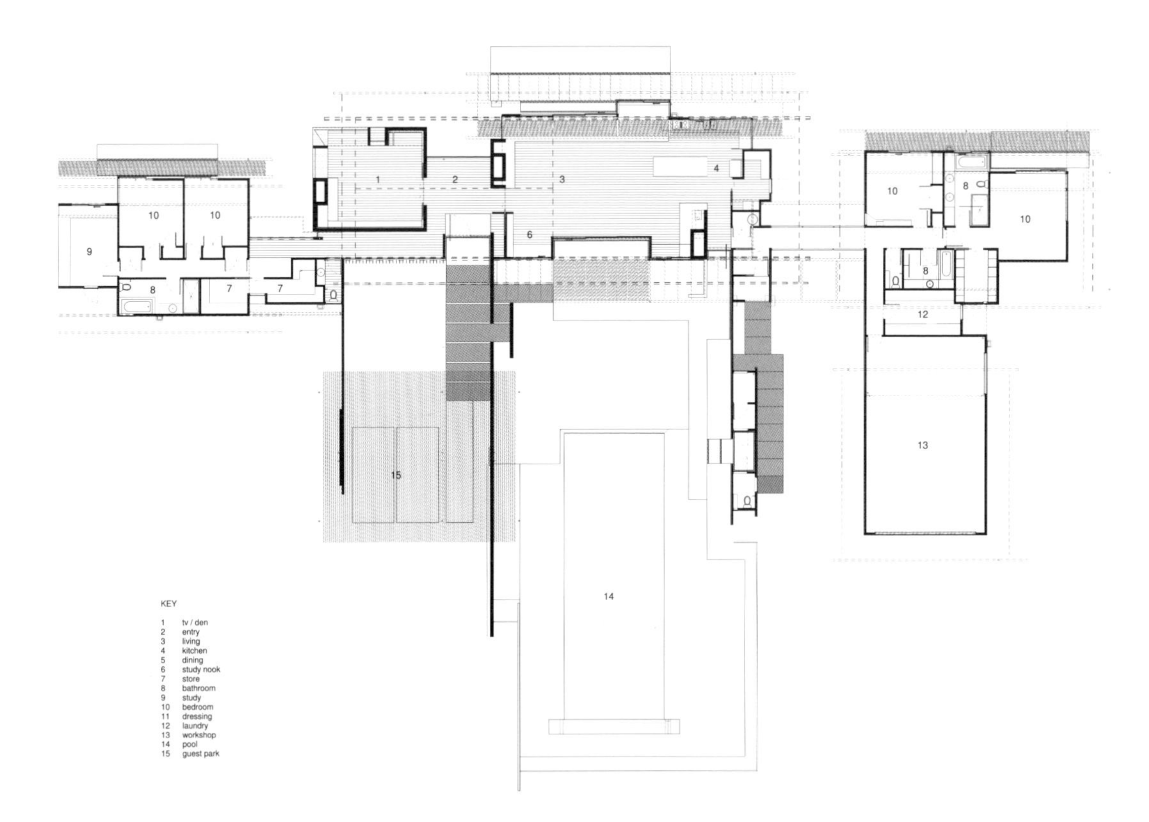

平面图 / Site Plan

图书在版编目（CIP）数据

家族荣耀　豪宅铭刻 / 黄滢 马勇 主编 . – 武汉 : 华中科技大学出版社 , 2014.9

ISBN 978-7-5680-0423-7

Ⅰ . ①家… Ⅱ . ①黄… ②马… Ⅲ . ①别墅 – 建筑设计 – 世界 – 图集 Ⅳ . ① TU241.1-64

中国版本图书馆 CIP 数据核字（2014）第 224897 号

家族荣耀　豪宅铭刻（上、下）　　黄滢 马勇 主编

出版发行：华中科技大学出版社（中国 · 武汉）
地　　址：武汉市武昌珞喻路 1037 号（邮编：430074）
出 版 人：阮海洪

责任编辑：熊纯　　责任监印：张贵君
责任校对：岑千秀　　装帧设计：筑美空间

印　　刷：利丰雅高印刷（深圳）有限公司
开　　本：889 mm × 1194 mm　1/12
印　　张：38（上册 20 印张，下册 18 印张）
字　　数：228 千字
版　　次：2015 年 01 月第 1 版 第 1 次印刷
定　　价：598.00 元（USD 119.99）

投稿热线：（020）36218949　　duanyy@hustp.com
本书若有印装质量问题，请向出版社营销中心调换
全国免费服务热线：400-6679-118 竭诚为您服务

they're little, girls dream about their perfect wedding

From the time they're little, girls dream about their perfect

From the time they're little, girls dream about their perfect

From the time they're little, girls dream about their perfect

My

swe

they're little, girls dream

From the time they're little, girls dream

From the time they're little, girls dream

From the time they're little, girls dream